How Much Inequality Is Fair?

# How Much Inequality Is Fair?

MATHEMATICAL PRINCIPLES OF A MORAL, OPTIMAL, AND STABLE CAPITALIST SOCIETY

## Venkat Venkatasubramanian

Columbia University Press
New York

Columbia University Press
*Publishers Since 1893*
New York   Chichester, West Sussex
cup.columbia.edu
Copyright © 2017 Columbia University Press
All rights reserved

Library of Congress Cataloging-in-Publication Data
Names: Venkatasubramanian, Venkataraman, 1955- author.
Title: How much inequality is fair? : mathematical principles of a moral, optimal, and stable capitalist society / Venkat Venkatasubramanian.
Description: New York : Columbia University Press, 2017. | Includes bibliographical references.
Identifiers: LCCN 2016046854 | ISBN 978-0-231-18072-6 (cloth : alk. paper) | ISBN 978-0-231-54322-4 (e-book)
Subjects: LCSH: Income distribution—Mathematical models. | Wealth—Mathematical models. | Equality—Mathematical models. | Capitalism—Moral and ethical aspects.
Classification: LCC HB523 .V46 2017 | DDC 330.1/60151—dc23
LC record available at https://lccn.loc.gov/2016046854

Columbia University Press books are printed on permanent and durable acid-free paper.
Printed in the United States of America

COVER DESIGN:
Fifth Letter

To my wondrous lineage of *Acharyas*, both spiritual and material, for all their generous *prasadhams*.

To Edwin T. Jaynes, whose pioneering contributions inspired this work.

To Indu, Govind, and Shriya for their love and support.

# Contents

| | | |
|---|---|---|
| List of Tables | | xi |
| List of Figures | | xiii |
| Preface | | xv |

**CHAPTER ONE**
**Extreme Inequality in Income and Wealth** — 1

| 1.1 | Rise of the Top 1% and Fall of the Bottom 90% | 1 |
|---|---|---|
| 1.2 | Executive Pay and High CEO Pay Ratios | 5 |
| 1.3 | Economic Inequality: What Do People Consider Fair? | 8 |
| 1.4 | Why Is Extreme Inequality Worrisome? | 12 |
| 1.5 | A New Mathematical Theory of Fairness in a Capitalist Society | 16 |

**CHAPTER TWO**
**Foundational Principles of a Fair Capitalist Society** — 25

| 2.1 | Philosophical Perspectives: Mill, Rawls, Nozick, and Dworkin | 25 |
|---|---|---|
| 2.2 | A System-Theoretic Perspective | 30 |
| 2.3 | What Is the Purpose of a Society? | 34 |
| 2.4 | BhuVai: A Hybrid Utopian Society of Universal Happiness | 38 |

**CHAPTER THREE**
**Distributive Justice in a Hybrid Utopia** — 43

| 3.1 | Income Distribution in BhuVai: Formulating the Problem | 44 |
|---|---|---|

| | | |
|---|---|---|
| 3.2 | A Microeconomic Game-Theoretic Framework: "Restless" Agents Model | 49 |
| 3.3 | Potential Game Formulation of Free-Market Dynamics | 55 |
| 3.4 | Is There an Equilibrium Income Distribution? | 61 |
| 3.5 | Connection with Statistical Thermodynamics | 62 |

## CHAPTER FOUR
### Statistical Thermodynamics and Equilibrium Distribution — 65

| | | |
|---|---|---|
| 4.1 | Brief History | 65 |
| 4.2 | "Molecular Society": Microstates, Macrostates, and Multiplicity | 67 |
| 4.3 | Phase Space, Entropy, and Statistical Equilibrium | 69 |

## CHAPTER FIVE
### Fairness in Income Distribution — 79

| | | |
|---|---|---|
| 5.1 | Complex Teleological Systems | 79 |
| 5.2 | Statistical Teleodynamics: Intrinsic and Extrinsic Properties | 81 |
| 5.3 | Phase Space in Statistical Teleodynamics | 82 |
| 5.4 | What Is the Equilibrium Income Distribution in BhuVai? | 84 |
| 5.5 | Is the Equilibrium Distribution Fair? | 86 |
| 5.6 | Entropy: Correcting the Misinterpretations | 102 |
| 5.7 | Replicator Dynamics | 112 |
| 5.8 | Is the Ideal Free Market Morally Justified? | 115 |
| 5.9 | Is the Equilibrium Income Distribution Stable? | 121 |
| 5.10 | Fairness, Stability, and Robustness of a Free-Market Society | 123 |
| 5.11 | Comparison with Walrasian General Equilibrium | 130 |
| 5.12 | Bipopulation Game | 135 |
| 5.13 | Midpoint Summary | 138 |

## CHAPTER SIX
### Global Trends in Income Inequality: Theory Versus Reality — 143

| | | |
|---|---|---|
| 6.1 | Comparing Theory with Reality | 144 |
| 6.2 | A New Measure of Fairness in Income Inequality: $\psi$ | 147 |

| | | |
|---|---|---|
| 6.3 | Scandinavia: Almost Utopian | 154 |
| 6.4 | Potential Loss of Growth due to Extreme Inequality | 155 |
| 6.5 | Are the Rich Different from You and Me? | 157 |
| 6.6 | Gini Coefficient versus $\psi$ | 158 |
| 6.7 | One-Class and Two-Class Societies: Simulation Results | 162 |
| 6.8 | Summary | 165 |

## CHAPTER SEVEN
### What Is Fair Pay for Executives? — 169

| | | |
|---|---|---|
| 7.1 | Theory's Predictions for Large Corporations | 171 |
| 7.2 | How Much of a Company's Success Is due to the Executives? | 175 |

## CHAPTER EIGHT
### Final Synthesis and Future Directions — 191

| | | |
|---|---|---|
| 8.1 | Mathematical Foundations of a Utopian Capitalist Society: Maximum Fairness as the Design Principle | 192 |
| 8.2 | Statistical Teleodynamics: Macroview | 201 |
| 8.3 | Statistical Teleodynamics: Microview | 208 |
| 8.4 | Significance of the Utility Function and Its Parameters | 212 |
| 8.5 | Can a Simple Model Handle Free-Market's Complexity? | 217 |
| 8.6 | What About Economic Growth? | 225 |
| 8.7 | Comparison with Econophysics | 227 |
| 8.8 | Future Directions: Theoretical and Empirical Studies | 230 |
| 8.9 | Summary and Final Thoughts | 234 |

| | |
|---|---|
| **Notes** | 243 |
| **Bibliography** | 251 |
| **Index** | 267 |

# Tables

| | | |
|---|---|---|
| 1.1 | CEO Pay Ratio: Actual, Estimated, and Ideal | 8 |
| 5.1 | Feasible Assignments and Their Multiplicities | 93 |
| 5.2 | Maximum Entropy Distributions | 110 |
| 6.1 | Model Predictions and Sample Empirical Data for Different Countries | 146 |
| 6.2 | Two-Class System Parameters | 163 |
| 7.1 | Comparing Top CEO Salaries with Minimum Annual Salary of $25K: Actual versus Ideal Ratios | 173 |

# Figures

| | | |
|---|---|---|
| 1.1 | Income inequality in the United States from 1910 to 2010 | 2 |
| 1.2 | Income share of top 1% in the United States from 1910 to 2010 | 3 |
| 1.3 | Income inequality in Anglo-Saxon countries | 4 |
| 1.4 | Income inequality in continental Europe and Japan | 4 |
| 1.5 | Wealth inequality in Europe versus the United States | 5 |
| 1.6 | CEO pay ratio trend in the United States | 6 |
| 1.7 | Relative preference for three different wealth distributions | 10 |
| 1.8 | Actual, estimated, and ideal U.S. wealth distributions | 11 |
| 1.9 | Actual, estimated, and ideal U.S. wealth distributions by income level, political affiliation, and gender | 12 |
| 3.1 | Spreading of the pay distribution under competition in an ideal free-market environment | 47 |
| 3.2 | Net utility: inverted-U curve | 53 |
| 4.1 | All gas molecules are in the left chamber | 72 |
| 4.2 | Statistical equilibrium: uniform distribution of gas molecules | 73 |
| 4.3 | An extremely unlikely configuration | 74 |
| 6.1 | Income inequality in Norway, 1930–2010 | 148 |
| 6.2 | Income inequality in Sweden, 1920–2010 | 148 |
| 6.3 | Income inequality in Denmark, 1920–2010 | 149 |
| 6.4 | Income inequality in Switzerland, 1930–2010 | 149 |
| 6.5 | Income inequality in the Netherlands, 1920–2000 | 150 |
| 6.6 | Income inequality in Australia, 1940–2010 | 150 |

| | | |
|---|---|---|
| 6.7 | Income inequality in France, 1920–2000s | 151 |
| 6.8 | Income inequality in Germany, 1950–2010 | 151 |
| 6.9 | Income inequality in Japan, 1940–2010 | 152 |
| 6.10 | Income inequality in Canada, 1940–2000 | 152 |
| 6.11 | Income inequality in the United Kingdom, 1930–2010 | 153 |
| 6.12 | Income inequality in the United States, 1930–2010s | 153 |
| 6.13a | Gini coefficients (OECD) | 159 |
| 6.13b | Gini coefficients (LIS) | 160 |
| 6.14 | Agent-based simulation results | 163 |
| 6.15 | Power law fits truncated lognormal | 165 |

# Preface

The political problem of mankind is to combine three things: economic efficiency, social justice, and individual liberty.

JOHN MAYNARD KEYNES

---

Philosophy is written in this grand book—I mean the universe—which stands continually open to our gaze, but it cannot be understood unless one first learns to comprehend the language in which it is written. It is written in the language of mathematics.

GALILEO GALILEI

ONE MIGHT WONDER WHAT a chemical engineer is doing writing a book on income inequality. The importance of the topic is obvious—it is the foundational challenge of our time. Like an iceberg looming out of a fog, extreme income inequality has the potential to disrupt the very functioning and stability of our democracy in a fundamental way. But why is a chemical engineer writing about it? This is a reasonable question that deserves an appropriate answer. First, while I am indeed a professor of chemical engineering, by inclination, education, and experience, I am more of a complex systems theorist, interested in understanding fundamental conceptual principles behind the organization and functioning of complex adaptive systems. In particular, I am interested in understanding the design, control, optimization, and risk management issues in self-organized complex adaptive systems. Hence my interest in studying a free-market society from that perspective, even though most of my past contributions have involved addressing such issues in the context of chemical process systems.

Second, my interest in this area started with a question I had asked myself, as a graduate student in chemical engineering, in 1983: What would be a statistical mechanics–like framework for predicting the macroscopic behavior of a large collection of intelligent agents? At that time, I was writing my doctoral thesis, which was on the application of statistical mechanics techniques for the prediction of vapor-liquid properties of polar mixtures. I was also getting interested in artificial intelligence (which I pursued subsequently upon graduation as a post-doctoral researcher in artificial intelligence [AI] at Carnegie Mellon University). So as I thought about AI entities and their behavior, this question arose in my mind.

As I couldn't find the framework I was looking for in the literature at that time, I went about developing one in the following years, not realizing it would take me three decades. Since I thought this was too risky a problem to work on full time, particularly as a young researcher trying to get tenure in chemical engineering and build a career, I diversified my research portfolio by also working on certain problems in process systems engineering. I worked on this question intermittently, whenever my main work permitted, typically over the summer and winter breaks. This question led me to learning about all kinds of topics that chemical engineers typically don't think about, such as economics, finance, game theory, and so on. I was having a lot of fun learning new things and reflecting about this question from different perspectives. I remember thinking, all along, that even if this quest went nowhere, which was a distinct possibility in my mind, the intellectual enjoyment I was having was well worth the wild goose chase.

Finally, after nearly two decades of almost no progress but lots of fun, I had my first break, in 2003, when I started thinking about the design of self-organizing adaptive complex networks for optimal performance in a given environment. Our paper[1] on this topic taught me how the microstructure of a network is related to its macroscopic properties, and how the environment plays a critical role in determining the optimal balance between efficiency and robustness trade-offs in design and operation. I also developed a good feel for, and insights about, self-organizing dynamics and the emergence of new system-level behaviors, which turned out to be crucial when I started thinking about Adam Smith's "invisible hand" and the dynamics of the free market five years later.

This work led me, in 2004, to investigate the maximum entropy framework in the design of complex networks,[2] which resulted in the formulation of my initial ideas about *statistical teleodynamics*[3] in 2007.[4] Soon I realized that the true essence of entropy is fairness in a distribution, which got me interested in fairness in income distribution.

This resulted in my papers in 2009[5] and 2010,[6] wherein I proved that the lognormal distribution is the fairest distribution of pay in an ideal free market at equilibrium, which is determined by maximizing entropy, i.e., by maximizing fairness. This was a statistical mechanics–information-theoretic perspective, and I knew that there should be a game-theoretic answer as well to this. So I started working on that right away, and when I moved to Columbia in 2011, I initiated a collaboration with Jay Sethuraman, a game theory expert. Since I felt this was a well-defined problem that could be solved in a matter of months, and not years as the early stages had been, I felt comfortable to engage my doctoral student Yu Luo, who was working on a game theoretic framework for regulating self-interested agents, as a part of our team. This resulted in our 2015 paper.[7]

Essentially, my long quest for a statistical mechanics–like framework for rational intelligent agents started in 1983 and ended with our 2015 paper, thirty-two years later. In that paper, however, I had not fully explored the connection between the philosophical theories of human societies and the statistical teleodynamics perspective, even though I had touched upon those connections in my 2009 and 2010 papers. This book is written to address that gap as well as to present all the results from my past papers, and some new results, in a unified conceptual framework.

As a proponent of a new theory, one runs the risk of becoming a casualty of the adage "No one believes in a theory except the one who proposed it, and everyone believes in an experiment except the one who performed it!" This is further compounded when one works in an interdisciplinary field, as I have, for one runs the risk of not being taken seriously by either camp. Given my theory's transdisciplinary nature—spanning all the way from game theory and statistical mechanics to economics and political philosophy—it may be ignored and banished to an intellectual no man's land. Political philosophers might discard it, saying it has too much economics and math, and not enough philosophy, to be of relevance to them. On the other hand, economists might reject it because it has too much statistical physics. The physicists, particularly econophysicists, might pay some attention, and for that I am very grateful, but they are not the intended audience of this theory. This book is really meant for economists and sociologists, for the problem I address, namely, distributive justice in a capitalist society, is a central problem in their domains.

As will be apparent from my approach and writing, it is the perspective of an outsider, one who has had little formal training in economics or in political philosophy. While the drawbacks of being an outsider are many, it is not, however, without some benefits. The main benefit is that I feel unburdened by tradition and orthodoxy in political economy, thereby raising questions that have not been asked before or considering conceptual frameworks that are far outside the mainstream. Besides a philosopher's, sociologist's, or economist's perspective on a free-market society, it is quite useful, as I demonstrate, to get a systems engineer's viewpoint as well, for, after all, a free-market economy is a dynamical system with millions of interacting agents exhibiting statistical behavior. However, one of the drawbacks of my outsider perspective is that the treatment is not as rigorous and comprehensive, philosophically and economically, as one might have liked.

While this book is not meant as a popular exposition aimed at the general public, I have deliberately adopted a narrative style that is less formal than what is usually expected in a research monograph to make my theory as accessible as possible to a wider audience, and not just professional economists. Some readers might also find my style to be somewhat repetitive, particularly regarding the key ideas that are reiterated several times throughout the book, like a musical motif. My apologies, but I am a firm believer in the admonition of the great mathematician David Hilbert to a young Hermann Weyl, his new teaching assistant, on instruction: "Five times, Hermann, *five* times!"

While we are on the topic of pedagogy, I would like to note that I have only used as much mathematics as necessary to explain my theory. I have avoided being too formal, with axioms, proofs, and lemmas, as is common in mathematical economics papers, to improve readability and accessibility. While I have made a conscious attempt to avoid too formal an approach, the extent of mathematics in this book cannot be avoided for, after all, the central contribution of my theory is the mathematical formalism. One might consider my theory as mathematization of the insights of Locke, Smith, Mill, Rawls, Nozick, and Dworkin in the context of distributive justice. My theory is an attempt in search of universal mathematical principles that might underlie the emergence of organized behavior in both inanimate and animate complex systems. This desire, perhaps misplaced, is reflected in the admittedly grandiose subtitle of this book. I have agonized greatly over this, and I continue to, but my earlier choices, while more modest, were too wonkish to be suitable, as a number of people correctly pointed out. I hope I may be forgiven for letting my sentiment get the better of me with the final choice I have made.

My theory is essentially a work of synthesis, integrating disparate concepts and techniques, even though it has significant analytical forays. As I went about developing this theory, I often felt more like an artist than an engineer or a scientist. I felt like a jeweler exploring an intellectual mine of gems and precious metals, hoping to find beautiful pieces that would neatly fit in the design of jewel I had in mind, wanting to create an ornament that is beautiful yet functional, one that people would find to be both elegant and effective in addressing questions in distributive justice. Of the gems I found, the most exquisite one, in my mind, is *entropy*, which is the centerpiece in my theory. Like the exceptionally beautiful Koh-i-Noor diamond, it illuminates this dark, labyrinthine, intellectual landscape, showing the path and the destination, to arrive at a theory that unifies several seemingly disparate concepts from different domains.

This theory obviously has many limitations, as I point out in the book. I have tried, deliberately, to keep the models as simple as possible without losing key insights and relevance to empirical results in economics. I'd argue that, despite its shortcomings, this theory opens up new vistas in economics, analytical sociology, and political philosophy, suggesting new directions of inquiry that have the potential to yield new and useful results. While there has been exciting recent progress on the empirical front, as seen in the outstanding contributions of Atkinson, Piketty, Saez, and others, one doesn't get the sense that the theoretical front is also making such ground-breaking progress, going beyond the transformational ideas from the 1970s and 1980s, in distributive justice.

I am mindful of Sigmund Freud's caution, as quoted by E. T. Jaynes,[8] about the typical reception to new ideas: "Every new idea in science must pass through three phases. In Phase 1 everybody says the man is crazy and the idea is all wrong. In Phase 2 they say that the idea is correct but of no importance. In Phase 3 they say that it is correct and important, but we knew it all along."

While I have not been be able to find confirmation that Freud indeed had said this, there is nevertheless a lot of truth to it. As readers would no doubt recognize, several elements of the proposed theory are not new and indeed have been around for a long time. However, I believe that this is the first time one has articulated the deep connection between the central concepts in political philosophy, economics, statistical mechanics, information theory, game theory, and systems engineering—particularly, the insight that entropy, as a measure of fairness in a distribution, plays a central role in the dynamics of a free market under ideal conditions. I think the result that maximizing entropy (i.e., fairness) leads to equality of welfare in our hybrid utopia, thereby providing the *moral justification* for the free market, is also a significant and surprising insight. Even more surprising is the empirical finding that the Scandinavians have come very close to this utopia in reality!

I am grateful to a number of people who have contributed in one way or the other to this work. First, I thank all my students, postdoctoral associates, and collaborators, both past and present, who helped me learn so much about the design, control, and optimization issues in systems engineering, both the theoretical and the practical aspects. I learned how complex systems are supposed to work in theory, and how they can fail in practice, sometimes systemically. I consider extreme economic inequality as a systemic failure of a free-market society.

I am particularly grateful to Santhoji Katare, Priyan Patkar, and Fang-Ping Mu, my former doctoral students at Purdue who worked with me on self-organizing, adaptive, optimal networks. I also owe a great debt to Jay Sethuraman and Yu Luo for their valuable contributions in our game theory work. I am thankful to Yu Luo, in addition, for his assistance with the figures and LaTeX formatting of this book. I very much appreciate the constructive criticisms provided by the anonymous reviewers of my 2009 and 2010 papers in *Entropy*, of our 2015 paper in *Physica*, as well as of the manuscript version of this book. Their valuable comments shaped my thinking and helped me develop a more comprehensive conceptual framework. My thanks are also to Andrew Hirsch, William Masters, Dimitris Politis, Rajiv Sethi, and Wally Tyner, who have provided valuable feedback on my earlier work on this topic.

I very much appreciate the help I received from Ted Bowen and Lavinia Lorch who read a draft version and suggested improvements.

It is with great pleasure, and a sense of deep gratitude, that I acknowledge the invaluable support and guidance of my editor, Bridget Flannery-McCoy, of Columbia University Press. Not too many editors would have taken my work seriously, an unsolicited manuscript on a most challenging topic in economics from an outsider, but she did. For that, I am indebted to her forever. It is also a pleasure to acknowledge the members of her editorial team, Ryan Groendyk, Marisa Lastres, Ben Kolstad, and Ruthie Whitehead, for their expert assistance in a number of areas, particularly in copyediting.

Last but not least, it is an immense pleasure to thank my teachers—K. C. E. Dorairaj, my freshman physics professor, for initiating me in a career of asking scientific questions; P. R. Subramaniam, my undergraduate physics professor, for kindling my research interest in physics further; Keith Gubbins, my doctoral adviser, for introducing me to statistical mechanics and for accommodating my varied interests; and William Streett, my doctoral committee member, for his belief in me.

In conclusion, I take some comfort in Blaise Pascal's remark,[9] "Let no one say that I have said nothing new... the arrangement of the subject is new." This is particularly relevant here because all the key concepts, such as entropy and potential, have been known for a long time. What is new, as noted, is their interpretation and their connection with the fundamental principles of political philosophy and free-market economics.

Therefore, I humbly submit this work for your consideration with considerable trepidation, yet with the hope that it will be given a fair hearing and that it might add, despite its many shortcomings, something useful to the vital subject of distributive justice.

CHAPTER ONE

# Extreme Inequality in Income and Wealth

Government of the people, by the people, for the people, shall not perish from the Earth.
ABRAHAM LINCOLN

---

What improves the circumstances of the greater part can never be regarded as an inconveniency to the whole. No society can surely be flourishing and happy, of which the far greater part of the members are poor and miserable.
ADAM SMITH

## 1.1  Rise of the Top 1% and Fall of the Bottom 90%

IN RECENT YEARS, there has been growing concern over the widening inequality in income and wealth distributions in the United States and elsewhere (Milanovic 2012; Stiglitz 2012; Piketty and Goldhammer 2014; Krugman 2014; Reich 2015; Stiglitz 2015; Krugman 2015; Atkinson 2015). The statistics are troubling. For instance, as of 2012, the top 1% of households held 41.8% of all wealth in the United States (Saez and Zucman 2014), up from a post–World War II low of about 20% in 1976 (Domhoff 2006). An important source of the wealth inequality is a similar trend in income distribution. Income remains highly concentrated. The top 1% of earners received 17.9% of all income in 2012 in the United States, an increase from 12.8% in 1982 (Domhoff 2006; Piketty and Saez 2003). Income is what people earn from work, but it also includes, if any, dividends, interest, and rents or royalties that are paid to them on assets they own.[1] Attendees of the 2015 World Economic Forum in Davos, Switzerland, identified income inequality as the most challenging global trend (Mohammed 2015). There is much discussion in both academic literature and the popular press about what this means, what the consequences are, and what can or should be done about it (Bebchuk and Fried 2005; Jones 2009b; Stiglitz 2012; Mishel et al. 2012; Klinger et al. 2013; Cassidy 2014b; Kristof 2014; Krugman 2014; Piketty and Goldhammer 2014; Reich 2015; Krugman 2015; Atkinson 2015; Mohammed 2015).

In *Capital in the Twenty-First Century*, economist Thomas Piketty demonstrates these global trends vividly (see figures 1.1 to 1.5). As *New Yorker* writer John Cassidy observes in his review of Piketty's 700-page tome (Cassidy 2014a, b), figure 1.1 shows the share of overall income taken by the top 10% of households from 1910 to 2010 in the United States. Generally speaking, the trend is U-shaped, where inequality climbed steeply in the Roaring Twenties and then fell sharply in the decade and a half following the Great Crash of October 1929. From the mid-1940s to the mid-1970s, for about 30 years, it stayed fairly stable, and then it started rising, eventually topping the 1928 level in 2007. It dropped somewhat after the financial crisis in 2007–2008, but by 2015 it had risen again to about 50%.

Figure 1.2 shows the share of income going to the top 1% over the same period. The top line, which includes income of all kinds, has the

*Figure 1.1* Income inequality in the United States from 1910 to 2010 from Piketty and Goldhammer (2014).

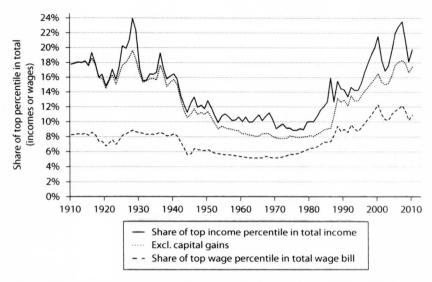

*Figure 1.2* Income share of top 1% in the United States from 1910 to 2010 from Piketty and Goldhammer (2014).

same U shape. The top percentile hasn't taken such a large share of overall income since 1928. However, the recent rise is less pronounced when one examines just the wage income (bottom line). The difference between the bottom line (wage income) and the top line (total income) is due to income from capital such as interest payments, dividends, and capital gains. Since the top 1% own a lot of wealth, they receive much of their income in this manner. Nonetheless, the gap in wage income has widened significantly in recent years.

Figures 1.3 and 1.4 confirm that this is a global phenomenon. We see it in Australia, Canada, France, Germany, Japan, Sweden, and the United Kingdom. However, the rise in the U curve is less pronounced in Sweden, France, Japan, Australia, and Germany. The United States has the most extreme income inequality now, but this wasn't the case from 1945 to 1975, when it was more equitable than Canada and on a par with France. The United States was a more economically just society then than it is now.

As noted, extreme inequalities exist in wealth as well. As figure 1.5 shows, the top 1% and top 10% control the lion's share in European and U.S. societies. For much of the nineteenth and twentieth centuries, Western Europe was dominated by an elite that possessed much of the

4  *Extreme Inequality in Income and Wealth*

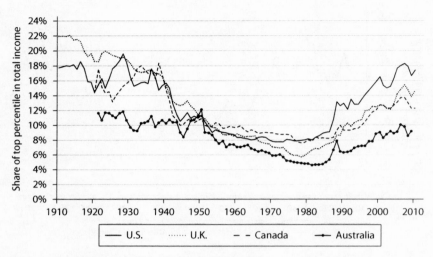

*Figure 1.3* Income inequality in Anglo-Saxon countries from Piketty and Goldhammer (2014).

land and the wealth. The United States also had rich and poor classes, but wealth was distributed a bit more widely. For example, in 1910, the top 10% in Europe owned about 90%; in the United States, the top 10% owned about 80%. However, in recent decades wealth has become more concentrated in the United States than it is in Europe. In 2010,

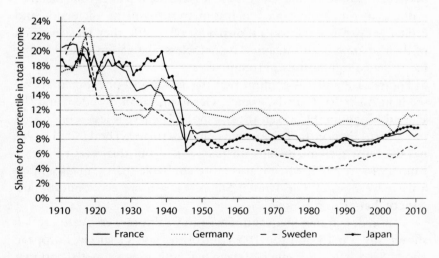

*Figure 1.4* Income inequality in continental Europe and Japan from Piketty and Goldhammer (2014).

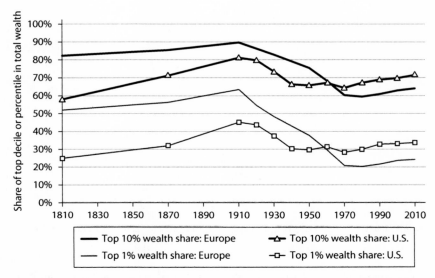

*Figure 1.5* Wealth inequality in Europe versus the United States from Piketty and Goldhammer (2014).

the American 1% owned about 33% of all wealth: the European 1% owned about 25%.

## 1.2 Executive Pay and High CEO Pay Ratios

A related trend of equally great concern is the excessive pay packages for CEOs, which are reflected in the extraordinarily high CEO pay ratios, particularly in the United States (Anderson et al. 2008; Hargreaves 2014; Holmberg and Schmitt 2014; Mishel and Davis 2014; Nocera 2014). The ratio of CEO salary (i.e., total compensation including bonuses and stock options) to that of an average employee has risen from the 25 to 40 range in the 1970s to more than 300 in recent years in the United States (see figure 1.6) (Mishel and Davis 2014).

These comparisons naturally raise the question: What is fair pay for a CEO relative to the pay of other employees?

A common response to this question is that the free market takes care of all this and determines the values of a CEO and the other employees. But is the market really efficient in determining CEO pay? As Robert Reich argues in his recent book *Saving Capitalism* (Reich 2015, xiv),

6  *Extreme Inequality in Income and Wealth*

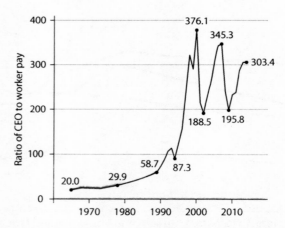

*Figure 1.6* CEO pay ratio trend in the United States (Mishel and Davis 2014).

"The meritocratic claim that people are paid what they are worth in the market is a tautology that begs the questions of how the market is organized and whether that organization is morally and economically defensible. In truth, income and wealth increasingly depend on who has the power to set the rules of the game." This important point is also emphasized by Lucian Bebchuk and Jesse Fried in their book *Pay without Performance: The Unfulfilled Promise of Executive Compensation* (Bebchuk and Fried 2006). The widely held assumption of arm's-length contracting in determining pay (i.e., contracting between executives attempting to get the best possible deal for themselves and boards trying to get the best deal for shareholders) is generally not upheld as much as one would like. As a result, managerial power has played a key role in sharply increasing executive pay.

Bebchuk and Fried (2005, 8–9) observe:

> The pervasive role of managerial power can explain much of the contemporary landscape of executive compensation, including practices and patterns that have long puzzled financial economists. We also show that managerial influence over the design of pay arrangements has produced considerable distortions in these arrangements, resulting in costs to investors and the economy. This influence has led to compensation schemes that weaken managers' incentives to increase firm value and even create incentives to take actions that reduce long-term firm value.

They further elaborate:

> Flawed compensation arrangements have not been limited to a small number of "bad apples"; they have been widespread, persistent, and systemic. Furthermore, the problems have not resulted from temporary mistakes or lapses of judgment that boards can be expected to correct on their own; rather they have stemmed from structural defects in the underlying governance structure that enable executives to exert considerable influence over their boards. The absence of effective arm's length dealing under today's system of corporate governance has been the primary source of problematic compensation arrangements. Finally, while recent reforms that seek to increase board independence will likely improve matters, they will not be sufficient to make boards adequately accountable; much more needs to be done.

Such a view is not held by academicians alone. The well-known compensation consultant Graef "Bud" Crystal (1991) wrote a highly critical commentary on the executive compensation practices in his book *In Search of Excess*. Things have worsened since then. Peter Drucker, the management guru, in his 2004 interview for *Fortune* was equally critical (Drucker and Schlender 2004):

> In every boom there is a tendency toward hero-worship of CEOs. The smart CEOs methodically build a management team around them. But many of those celebrity CEOs who are so highly regarded don't know what a team is. Moreover, the compensation inflation for CEOs has done very real damage to the concept of the management team. In an executive program I have at Claremont, the typical students are general managers of major divisions at very large companies, and they are very well paid. But it's fair to say they are contemptuous of the excessive pay that many of their CEOs receive. J. P. Morgan once said the top manager of a company should have a salary 20 times that of the rank-and-file worker. Today it is more like 400 times that. I'm not talking about the bitter feelings of the people on the plant floor. They're convinced that their bosses are crooks anyway. It's the midlevel management that is incredibly disillusioned. And so the present crisis of the CEO is a serious disaster. Let me again quote J. P. Morgan, who said, "The CEO is just a hired hand."

## 1.3 Economic Inequality: What Do People Consider Fair?

What, then, is fair compensation for CEOs? What do people think it should be? Do people even *care* about fairness?

Recently, business school professors Sorapop Kiatpongsan and Michael Norton (2014) asked 55,000 people in forty countries (including about 1600 people in the United States) these questions. A nice summary of their work can be found in Weissmann (2014), and we paraphrase it here. They asked participants how much they thought CEOs made compared with the average low-skill factory worker, and how much they should make. Table 1.1 shows the estimated, ideal, and actual ratios.

They found that there is *remarkable consistency* across countries and demographics. For example, Americans believed 6.7:1 would be ideal, while Australians said 8.3:1; the French settled at 6.7:1 and the Germans around 6.3:1. Contrary to cultural stereotypes, Americans and Europeans hold similar notions of fairness. The Scandinavians arrived at a lower value around 2. But the entire range is about 2 to 8.

TABLE 1.1
CEO Pay Ratio: Actual, Estimated, and Ideal

| Country | Actual ratio of CEO to worker pay | What subjects thought was the ratio | What subjects said would be the ideal ratio |
| --- | --- | --- | --- |
| Australia | 93 | 40 | 8.3 |
| Austria | 36 | 12 | 5 |
| Czech Republic | 110 | 9.4 | 4.2 |
| Denmark | 48 | 3.7 | 2 |
| France | 104 | 24.2 | 6.7 |
| Germany | 147 | 16.7 | 6.3 |
| Israel | 76 | 7 | 3.6 |
| Japan | 67 | 10 | 6 |
| Norway | 58 | 4.3 | 2.3 |
| Poland | 28 | 13.3 | 5 |
| Portugal | 53 | 14 | 5 |
| Spain | 127 | 6.7 | 3 |
| Sweden | 89 | 4.4 | 2.2 |
| Switzerland | 148 | 12.3 | 5 |
| United Kingdom | 84 | 13.5 | 5.3 |
| United States | 354 | 29.6 | 6.7 |

The data also revealed that people *dramatically underestimated* actual pay inequality. In the United States, the underestimation was particularly pronounced. The actual pay ratio of CEOs to unskilled workers (354:1) far exceeded the estimated ratio (30:1), which in turn exceeded the ideal ratio (7:1). In general, respondents underestimated actual pay gaps, and their ideal pay gaps are even further from reality than those underestimates.

Earlier in 2011, Michael Norton and Dan Ariely conducted a similar study regarding the distribution of wealth. They conducted two experiments by surveying about 5500 people, randomly drawn from a panel of more than 1 million Americans. In the first, participants were shown three unlabeled pie charts that depict possible wealth distributions, as shown in figure 1.7. Each slice depicts the percentage of wealth possessed by each quintile of the population: one that was totally equal, one based on Sweden's highly egalitarian income distribution, and one based on the distribution of wealth in the United States, which is significantly skewed toward the top. They presented respondents with the three pairwise combinations of these pie charts (in random order) and asked them to choose which nation they would rather live in, given a Rawlsian "veil of ignorance" situation. Norton and Ariely (2011) wrote, "In considering this question, imagine that if you joined this nation, you would be randomly assigned to a place in the distribution, so you could end up anywhere in this distribution, from the very richest to the very poorest."

As can be seen in figure 1.7, the (unlabeled) U.S. distribution was far less desirable than both the (unlabeled) Swedish distribution and the equal distribution, with some 92% of Americans preferring the Swedish distribution to that of the United States. In addition, this overwhelming preference for the Swedish distribution over the U.S. distribution was robust across gender (females 92.7%, males 90.6%), preferred candidate in the 2004 election (Bush voters 90.2%, Kerry voters 93.5%), and income (less than $50,000: 92.1%; $50,001 to $100,000: 91.7%; more than $100,000: 89.1%). In addition, there was a slight preference for the distribution that resembled that of Sweden (51%) relative to the equal distribution (49%), suggesting that *Americans prefer some inequality to perfect equality*, but *not to the degree* currently present in the United States.

10    Extreme Inequality in Income and Wealth

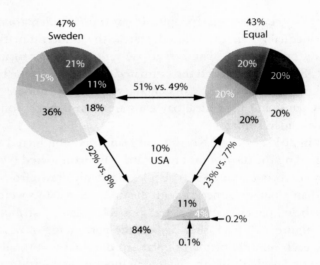

*Figure 1.7* Relative preference among all respondents for three distributions: Sweden (upper left), an equal distribution (upper right), and the United States (bottom). Pie charts depict the percentage of wealth possessed by each quintile; for instance, in the United States, the top wealth quintile owns 84% of the total wealth, the second-highest 11%, and so on (Norton and Ariely 2011).

In the second experiment, Norton and Ariely asked participants to guess how wealth was distributed in the United States and then to specify how it would be shared in an ideal world. It turns out that Americans have little idea how concentrated wealth truly is.

The participants estimated that the top 20% of U.S. households have about 59% of the country's net worth, whereas in the real world they have about 84% of it. In their utopia, they said that the top quintile would control 32% of the wealth (figure 1.8).

Both studies—the wealth inequality in 2011 and the income inequality in 2014—suggest three main messages, as the authors observe. First, most Americans (in fact, most people in the world) drastically underestimate the current disparities in wealth and income in their society. Second, Americans seem to prefer to live in a country more like Sweden than like the United States. Third, there is much greater consensus than disagreement across groups from different sides of the political spectrum about this desire for a more equal distribution of wealth

Extreme Inequality in Income and Wealth 11

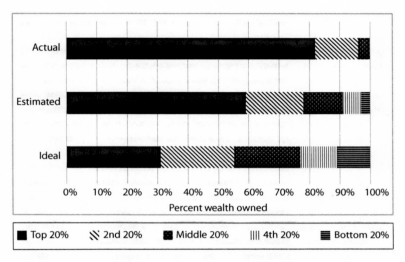

*Figure 1.8* Actual U.S. wealth distribution versus estimated and ideal distributions across all respondents. Because of their small percentage share of total wealth, both the "4th 20%" value (0.2%) and the "bottom 20%" value (0.1%) are not visible in the actual distribution (Norton and Ariely 2011).

(figure 1.9). In fact, in this regard Americans are not all that different from their European counterparts.

Writing about the Norton and Ariely wealth study, Chrystia Freeland, then global editor-at-large at Thomson Reuters, quipped (Freeland 2011), "Americans actually live in Russia, although they think they live in Sweden. And they would like to live on a kibbutz." It's an elegant summary of the state of affairs in the United States.

There are a number of reasons why Americans have allowed the inequality to become so excessive, but one of the main ones is that people are not even aware of how excessive it really is. They have a much more naive impression. They may also hold overly optimistic beliefs about opportunities for social mobility in the United States, which in turn may drive support for unequal distributions of wealth (Norton and Ariely 2011; Keister 2005; Charles and Hurst 2003; Benabou and Ok 2001). Freeland (2011) writes about this "lottery effect," i.e., Americans are hopeful that someday they will win this lottery and get wildly rich as others have.

12   *Extreme Inequality in Income and Wealth*

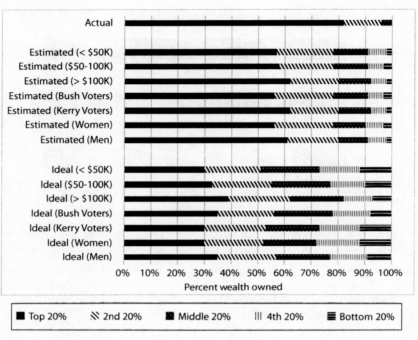

*Figure 1.9* Actual U.S. wealth distribution versus estimated and ideal distributions of respondents from different income levels, political affiliations, and genders. Because of their small percentage share of total wealth, both the fourth 20% value (0.2%) and the bottom 20% value (0.1%) are not visible in the actual distribution (Norton and Ariely 2011).

## 1.4   Why Is Extreme Inequality Worrisome?

Income inequality per se is not necessarily bad. Typically, in a society, different people have different talents, different skills, and different capacities to work, and hence they end up making different contributions, some more, others less. Therefore, it is only fair that those who contribute more earn more. We also *need* this inequality to incentivize talented, hard-working individuals to contribute more to society so that everyone can benefit from that value creation. Therefore, a certain degree of inequality makes sense not only at the individual level but also for the entire society.

But when a large section of the society is not benefiting from the society's growth (e.g., in GDP), with its real incomes stagnating for decades and its access to health care, education, and opportunities for upward mobility declining, then the system is not working as well as it should. There is increasing evidence that this has been happening to the bottom ~90% in the United States for the past three decades or so (Reich 2015; Stiglitz 2015). This extreme inequality in the United States, and elsewhere, is deeply troubling on a number of fronts.

First, there is the *moral* issue. For a country explicitly founded on the principles of *liberty*, *equality*, and the *pursuit of happiness*, protected by the "government of the people, by the people, for the people," the extreme inequality raises troubling questions of social justice that get at the very foundations of our society. Why and how did we become the "government of the 1% by the 1% for the 1%," as the Nobel Laureate in Economics Joseph Stiglitz (2011) wrote in his *Vanity Fair* essay? Stiglitz highlights these thoughts:

> Of all the costs imposed on our society by the top 1 percent, perhaps the greatest is this: the erosion of our sense of identity, in which fair play, equality of opportunity, and a sense of community are so important. America has long prided itself on being a fair society, where everyone has an equal chance of getting ahead, but the statistics suggest otherwise: the chances of a poor citizen, or even a middle-class citizen, making it to the top in America are smaller than in many countries of Europe.

Again, from the perspective of the foundational principles of our democratic capitalist society, Harvard philosopher Tim Scanlon (2005) argues that extreme inequality is bad for the following reasons: (1) economic inequality can give wealthier people an unacceptable degree of control over the lives of others; (2) economic inequality can undermine the fairness of political institutions; (3) economic inequality undermines the fairness of the economic system itself; and (4) workers, as participants in a scheme of cooperation that produces national income, have a claim to a fair share of what they have helped to produce. Scanlon argues:

> But the objections to inequality that I have listed rest on a different moral relation. It's the relation between individuals who are participants in a cooperative scheme. Those who are related to us in this way matter morally in a further sense: they are fellow participants to whom the terms of our cooperation must be justifiable.

In our current environment of growing inequality, can such a justification be given? No one has reason to accept a scheme of cooperation that places their lives under the control of others, that deprives them of meaningful political participation, that deprives their children of the opportunity to qualify for better jobs, and that deprives them of a share in the wealth they help to produce.

These are not just objections to inequality and its consequences: they are at the same time challenges to the legitimacy of the system itself. The holdings of the rich are not legitimate if they are acquired through competition from which others are excluded, and made possible by laws that are shaped by the rich for the benefit of the rich. In these ways, economic inequality can undermine the conditions of its own legitimacy.

As Scanlon argues, why should the bottom 90% play ball when they feel that the game is rigged to benefit mostly the top 10% or the top 1%? Robert Reich (2015, 163–65) elaborates the serious consequences of this important point in his book:

> If the game is perceived to be rigged in favor of those at the top, then others are more likely to view cheating as acceptable—stealing and pilfering from employers, rigging time clocks, overbilling, skimming off some of the profits, accepting small bribes and kickbacks.... But an economy is based on trust. The cumulative effect of even small violations of trust can generate huge costs.

He further observes:

> Second, when the game seems rigged and trust deteriorates, loyalty can no longer be assumed.... Employees or contractors withhold technical information or economic insights that, if shared, could boost joint productivity but could just as easily line the pockets of top executives and reduce the number of jobs.

Reich concludes with a dire warning:

> Third and finally, people who feel subjected to what they consider to be rigged games often choose to subvert the system in ways that cause everyone to lose.... In summary, when people feel that the system is unfair and arbitrary and that hard work does not pay off, we

all end up losing.... Together, these responses impose incalculable damage on an economic system. They turn an economy and a society into what mathematicians would call a "negative sum" game. When capitalism ceases to deliver economic gains to the majority, it eventually stops delivering them at all—even to a wealthy minority at the top. It is unfortunate that few of those at the top have yet to come to understand this fundamental truth.

Thus, the concerns are not only about the immorality and illegitimacy of extreme inequality, but also about its adverse effects on a range of outcomes we value, such as education, health care, economic growth, justice, opportunity for upward mobility, well-being and happiness, and more (Stiglitz 2015; Krugman 2014, 2015).

There is also the question of whether a capitalist society is *stable* under extreme inequality. As Stiglitz cautions,

> It is this sense of an unjust system without opportunity that has given rise to the conflagrations in the Middle East.... The top 1 percent have the best houses, the best educations, the best doctors, and the best lifestyles, but there is one thing that money doesn't seem to have bought: an understanding that their fate is bound up with how the other 99% live. Throughout history, this is something that the top 1 percent eventually do learn. Too late.

Reich (2015, xii) also stresses this warning: "The threat to capitalism is no longer communism or fascism but a steady undermining of the trust modern societies need for growth and stability. When most people stop believing they and their children have a fair chance to make it, the tacit social contract societies rely on for voluntary cooperation begins to unravel."

All this is not a matter of academic interest alone or a matter of concern for only the bottom 90%. Even the rich are beginning to get worried, as Peter Georgescu (2015) recently wrote in his op-ed piece in the *New York Times*: "I'm scared. The billionaire hedge funder Paul Tudor Jones is scared. My friend Ken Langone, a founder of the Home Depot, is scared. So are many other chief executives. Not of Al Qaeda, or the vicious Islamic State or some other evolving radical group from the Middle East, Africa or Asia. We are afraid where income inequality will lead."

Few observers are saying this is going to happen anytime soon, but this is certainly a path we don't want to be on down the line. Some roads are best left not taken, with all deference to Robert Frost!

To be fair, not all economists—even those who agree that extreme inequality is a bad thing—are convinced about the magnitude of the adverse impact. Lane Kenworthy (2015) comments:

> Let's step back and take stock. I've looked at the experiences of the world's rich countries in the period from 1979 to 2007 to see what they tell us about income inequality's effects on an array of social, economic, and political outcomes. My conclusion is that the available evidence suggests, first, that income inequality has done significant harm and, second, that inequality's harm has been less pervasive and devastating than some claim.

That said, Kenworthy does conclude in his paper that we should worry about the high and rising inequality in the United States.

To summarize, the objections raised against extreme inequality, the unfairness of it, rest primarily on three philosophical grounds, which we will explore in greater detail in chapter 2. From the utilitarian perspective, it reduces the overall welfare, happiness, or utility of a society as income and wealth get concentrated in the hands of a few. From the libertarian perspective, the resulting lack of economic freedom reduces other kinds of freedoms as well, thereby diminishing the rights and liberties of an average citizen in the society. Finally, from the moral perspective, it undermines the very social contract of a democratic capitalist society by breaking the implicit covenant of shared prosperity.

## 1.5 A New Mathematical Theory of Fairness in a Capitalist Society

Clearly, fairness in income and wealth distributions is a matter of utmost importance for a number of reasons: morality, growth, efficiency, stability, opportunity for upward mobility, access to health care, education, and justice, and so on. People do care a great deal about fairness. It is, after all, at the very core of a capitalist democracy.

But, as noted, different people have different talents, thereby making different contributions in a society. Therefore, people expect them to be compensated differently, commensurate with their contributions. So we naturally expect an unequal distribution of income and wealth in a society. Again, this is only fair as people who contribute more should earn more.

But how much more?

In other words, at the risk of sounding oxymoronic, *what is the fairest inequality of income?*[2] This question is at the heart of the inequality debate. The debate is not so much about inequality per se as it is about fairness. This is the central question we address in our theory.

As it turns out, there is no answer to this fundamental question in conventional economic theories. The common answer is that whatever income inequality the free-market mechanism delivers in practice is fair. There is extensive empirical literature on income and wealth distributions, and we cite only a sample here (Champernowne 1953; Champernowne and Cowell 1998; Milanovic 2002; Willis and Mimkes 2004; Bebchuk and Fried 2005; Kato and Kubo 2006; Anderson et al. 2008; Mishel et al. 2012; Piketty and Goldhammer 2014; Saez and Zucman 2014). Empirical observations are obviously very important, but it is equally important to complement them with a theoretical framework that provides a deeper understanding and analytical insight. As we described above, we can see *intuitively* that the current level of inequality is unfair, is morally and economically unjustified, and is potentially destabilizing. However, can we convert this intuitive, qualitative understanding into a more formal quantitative theory? That is, can we address this intuition mathematically and develop an analytical framework that can model and explore this understanding in more precise, quantitative terms? This is what we are after in this book.

### 1.5.1  Income Distribution: Four Fundamental Questions

In particular, from a theoretical perspective, we would like two sets of fundamental questions answered in mathematical and quantitative terms. (1) What kind of income distribution should arise, naturally, under ideal conditions, in a free-market environment comprising utility-maximizing employees and profit-maximizing companies? (2) Is this distribution fair?

The second set is related to the morality and stability concerns raised above. Now, morality obviously is a tricky concept as it can mean different things to different people in different contexts. The dictionary definition of morality is "a system or a set of rules about right versus wrong conduct." In our present context, it is not right that a large majority of the working population is struggling to make ends meet while a small minority is enjoying the fruits that the majority helped produce. Here morality is simply about whether the free-market system works for most people. On this basis, socialists and communists have argued that the free market is *immoral* because the benefits are reaped by a select few while the majority toil away. They argue that the free marketers' claim that "a rising tide lifts all boats"[3] is more like "a rising tide lifts all yachts" (Buren 2014) in practice. Recent trends in income and wealth inequality seem to support their argument.

Free-market enthusiasts, on the other hand, defend their system by arguing that the quality of life improvements made over the last two centuries in their societies (mostly Western democracies), pulling hundreds of millions of people out of poverty, are much greater than what has been accomplished in the socialist and communist societies. Their justification is that free-market capitalism may be immoral in theory, but in practice it works better than the socialist and communist systems, which are supposedly moral in theory but not so much in practice. So their defense is often based more on practical outcomes, on economic growth and efficiency, than on principle per se.

Adam Smith claimed that by letting every individual pursue his selfish drive to maximize his utility, the society also will benefit as an emergent phenomenon, even though it was not originally intended by the individual agents, through the action of the "invisible hand."

It is important at this juncture to explain a bit about the background and use of the famous "invisible hand." Smith used it three times in his two books *The Theory of Moral Sentiments* (Smith 1976) and *An Inquiry into the Causes of the Wealth of Nations* (1976).

The first use of this term occurs in *The Theory of Moral Sentiments*, in part 4, chapter 1, where Smith describes a selfish landlord as being led by an "invisible hand" to distribute his harvest among his serfs:

> The proud and unfeeling landlord views his extensive fields, and without a thought for the wants of his brethren, in imagination consumes himself the whole harvest.... [Yet] the capacity of his stomach bears no proportion to the immensity of his desires ... the rest he will be obliged to distribute among those, who prepare, in the nicest

manner, that little which he himself makes use of, among those who fit up the palace in which this little is to be consumed, among those who provide and keep in order all the different baubles and trinkets which are employed in the economy of greatness; all of whom thus derive from his luxury and caprice, that share of the necessaries of life, which they would in vain have expected from his humanity or his justice.... The rich ... are led by an invisible hand to make nearly the same distribution of the necessaries of life, which would have been made, had the earth been divided into equal portions among all its inhabitants, and thus without intending it, without knowing it, advance the interest of the society.

In the *Wealth of Nations*, Smith uses the metaphor in Book 4, chapter 2, paragraph 9:

But the annual revenue of every society is always precisely equal to the exchangeable value of the whole annual produce of its industry, or rather is precisely the same thing with that exchangeable value. As every individual, therefore, endeavours as much as he can both to employ his capital in the support of domestic industry, and so to direct that industry that its produce may be of the greatest value, every individual necessarily labours to render the annual revenue of the society as great as he can. He generally, indeed, neither intends to promote the public interest, nor knows how much he is promoting it. By preferring the support of domestic to that of foreign industry, he intends only his own security; and by directing that industry in such a manner as its produce may be of the greatest value, he intends only his own gain, and he is in this, as in many other cases, led by an invisible hand to promote an end which was no part of his intention. Nor is it always the worse for the society that it was not part of it. By pursuing his own interest he frequently promotes that of the society more effectually than when he really intends to promote it. I have never known much good done by those who affected to trade for the public good. It is an affectation, indeed, not very common among merchants, and very few words need be employed in dissuading them from it.

In other words, Smith and his followers reasoned intuitively that the ultimate outcome is beneficial to the entire society, increasing the total utility of the population. This outcome, in their view, provides the moral justification.

Again, we are seeking a mathematical basis for this intuitive reasoning in our theory. We believe it is important and useful to answer a second set of questions from a theoretical perspective: (3) Is a free-market society morally justified, at least under ideal conditions? (4) Is a free-market society stable, at least under ideal conditions?

The answers to these questions can serve as a fundamental benchmark against which we can evaluate the income distributions seen in real life. In the absence of such a reference framework, the conclusions we reach by relying on empirical observations alone are likely to be incomplete, in an important manner. This benchmark can help us measure and better understand the deviations caused by nonidealities in the real world, and to develop appropriate policy frameworks and incentive structures to try to correct the inequalities. It can give us a quantitative basis for understanding and developing rational tax policies, pay packages for executives, and so on. This is what we accomplish with our normative theory.

### 1.5.2 Econophysics Models: Conceptual Difficulties

As noted, there are no answers to these questions in conventional economic theories and models. When one explores outside mainstream economics, one finds that there has been much work, in the past two decades or so, in the econophysics community addressing the income distribution question. Econophysics researchers have modeled income and wealth distributions by applying concepts and techniques from statistical mechanics (Foley 1994; Levy and Solomon 1996; Foley 1996; Stanley et al. 1999; Foley 1999; Bouchaud and Mézard 2000; Dragulescu and Yakovenko 2000; Souma 2001; Willis and Mimkes 2004; Farmer et al. 2005; Richmond et al. 2006; Chatterjee et al. 2007, 2005; Smith and Foley 2008; Yakovenko and Rosser Jr. 2009; Foley 2010; Willis 2011; Yakovenko 2012; Chakrabarti et al. 2013; Cho 2014; Axtell 2014).

While these models are quite interesting and instructive, and they represent important progress in the field, they haven't bridged the rather wide conceptual gulf that exists between economics and econophysics (Gallegati et al. 2006; Ormerod 2010), particularly in two crucial areas. One, the typical particle model of agent behavior in econophysics, assumes agents to have nearly "zero intelligence," acting at random, with no intent or purpose. This does not sit well with an extensive body of economic literature spanning several decades, where one models, in the ideal case, a perfectly rational agent whose goal is to maximize its utility or profit by acting strategically, not randomly. So, from the perspective of an economist, it is quite reasonable to ask: How can theories and models based on the collective behavior of purpose-free, random molecules explain the collective behavior of goal-driven, optimizing, strategizing men and women?

Another major conceptual stumbling block is the role of entropy in economics. As Mirowski (1989, 389) observes: "The real impediments to the neoclassical embrace of thermodynamics are the concept of entropy and the implications of the second law." In statistical thermodynamics, equilibrium is reached when entropy—a measure of randomness or uncertainty—is maximized. So an economist might wonder why maximizing randomness or uncertainty would be helpful in economic systems. We all know that markets are stable, and function well, when things are orderly, with less uncertainty, not more.

This has led to an uneasy relationship with entropy in economics, typically ranging from grudging tolerance to outright rejection, as seen from the remarks of two Nobel Laureates in Economics, Amartya Sen and Paul Samuelson. Sen observed (Sen and Foster 1997, 35–36), while commenting on the Theil index, an entropy-based measure of inequality:

> Given the association of doom with entropy in the context of thermodynamics, it may take a little time to get used to entropy as a good thing ("How grand, entropy is on the increase!"), but it is clear that Theil's ingenious measure has much to be commended.... But the fact remains that it is an arbitrary formula, and the average of the logarithms of the reciprocals of income shares weighted by income shares is not a measure that is exactly overflowing with intuitive sense. It is, however, interesting that the concept of entropy used in the natural sciences can provide a measure of inequality that is not immediately dismissible, however arbitrary it may be.

Paul Samuelson was caustic and dismissive outright (Samuelson 1990, 256): "As will become apparent, I have limited tolerance for the perpetual attempts to fabricate for economics concepts of 'entropy' imported from the physical sciences or constructed by analogy to Clausius-Boltzmann magnitudes." In his Nobel Lecture, Samuelson (1972b, 254) said: "There is really nothing more pathetic than to have an economist or a retired engineer try to force analogies between the concepts of physics and the concepts of economics. How many dreary papers have I had to referee in which the author is looking for something that corresponds to entropy or to one or another form of energy?"

On yet another occasion he issued this warning (Samuelson 1972a, 450): "And I may add that the sign of a crank or half-baked speculator in the social sciences is his search for something in the social system that corresponds to the physicist's notion of 'entropy.'" These major conceptual hurdles have stranded neoclassical economics in the metaphors of classical mechanics for over a century, rather than it benefiting from the concepts and techniques of statistical thermodynamics, which is a more appropriate analogue. The typical econophysics papers that use statistical mechanics-based approaches to economics have not addressed these central conceptual challenges.

In addition to these conceptual challenges, there is a technical one due to the nature of the data sets in economics. As Ormerod (2010) and Perline (2005) discuss, one can easily misinterpret data from lognormal distributions, particularly from truncated data sets, as the inverse power law or other distributions. Therefore, empirical verification of econophysics models is still in the early stages.

Furthermore, most critically, econophysics models have not addressed the issues of fairness and morality.

### 1.5.3 Outline of the New Quantitative Theory of Fairness

Thus, there is no theoretical guidance on the issue of fairness in a freemarket society, from either mainstream economics or econophysics. However, as it turns out, we do have excellent guidance from political philosophers such as Rawls (1971), Nozick (1974), Dworkin (2001), Cohen (2008), and Nussbaum (2011). But their contributions are qualitative in nature, and hence they don't offer quantitative guidance that can answer the questions we raised above. Furthermore, they proposed societies that were rather pure in their characteristic political philosophy, whether it was egalitarianism (à la Rawls or Dworkin) or libertarianism (à la Nozick), whereas real-life societies are generally hybrids, mixing and matching different philosophies to deal with political realities.

To make matters even more complex, real-life societies are also dynamical systems where millions of citizens interact with one another and the environment in myriad ways. This critical feature has not been accounted for in the aforementioned philosophical theories. This perspective is also missing in mainstream economics. The dominant modeling paradigm in economics, with its use of concepts like forces (e.g., supply and demand forces), static equilibrium, elasticity, etc., is classical mechanics. The many-agent dynamical aspects of an economic system require an entirely different perspective. To borrow, once again, from physics, the closest analog would be statistical mechanics augmented by information theory.

To address the questions we raised above, we need a quantitative, analytical theory of fairness whose predictions can be verified with real-world data. This is the challenge we have addressed in our theory—this is what this book is about. Addressing this challenge required developing a rather unorthodox approach, a grand synthesis of foundational concepts from disparate disciplines into a unified theoretical framework that includes the key perspectives on this problem—the perspectives of political philosophy, economics, statistical mechanics, information theory, game theory, and systems engineering. This problem is indeed like the one in the six blind men and the elephant fable from Indian philosophy. This new paradigm requires integrating liberty, equality, and fairness principles from political philosophy with utility maximization by rational self-interested agents in economics, which dynamically interact with one another and the environment modeled by entropy and potential from statistical mechanics, information theory, and game theory, subject to the efficiency and robustness criteria from systems engineering.

The new framework unifies the concept of entropy from statistical mechanics and information theory with the concept of potential from game theory, and it proves these represent the concept of fairness in economics and philosophy. This deep insight is one of the two key concepts in our unified framework, which we call statistical teleodynamics. The other key insight is that when one maximizes fairness, all agents enjoy the same level of utility or economic welfare at equilibrium in an ideal free-market society. This result mathematically justifies the moral basis of an ideal free-market society. We also prove that the fairest inequality of pay is a lognormal distribution, at equilibrium, under ideal conditions. This turns out to be a very valuable reference state for evaluating real-life income distributions.

I have described different elements of this theory in my six earlier papers (Venkatasubramanian et al. 2004, 2006; Venkatasubramanian 2007, 2009, 2010; Venkatasubramanian et al. 2015). In those papers, I had not fully explored the connection between the philosophical theories of human societies and the statistical teleodynamics perspective, even though I touched upon those connections. This book is written to address that gap as well as to present all the results from my past work in a unified conceptual framework.

The book is organized as follows. In chapter 2 we discuss the philosophical theories on fairness, mainly the theories of Mill, Rawls, Nozick, and Dworkin. Then we motivate our system-theoretic framework for modeling hybrid societies and introduce a hybrid utopia, called *BhuVai*. In chapter 3 we develop the mathematical framework for distributive justice in BhuVai. In this chapter, we derive the mathematical criterion for the equilibrium income distribution that arises naturally in a free-market society under ideal conditions. In chapter 4 we provide an intuitive introduction to statistical mechanics as well as the concept of entropy and statistical equilibrium for nonexperts. In chapter 5 we discuss the deep connection between the game-theoretic formulation and statistical mechanics by developing the critical insight that entropy really is a measure of fairness in a distribution. This insight helps us synthesize fundamental concepts from philosophy, economics, game theory, statistical mechanics, information theory, and systems engineering into a coherent formalism, namely, *statistical teleodynamics*. In chapter 6 we compare the theory's predictions with empirical data on global income distributions as well as with simulations. We define a new measure of income inequality, one that measures the level of fairness (or unfairness) in the income distribution, and for that reason we think it is better than the Gini coefficient. We also demonstrate surprising insights revealed by our theory regarding the performance of Scandinavia and the United States as free-market societies. In chapter 7 we compare the theory's predictions regarding CEO pay with reality as well as with what people and experts consider as fair. In chapter 8, we summarize, clarify certain subtle issues, and offer some thoughts on future directions in theoretical and empirical research.

CHAPTER TWO

# Foundational Principles of a Fair Capitalist Society

Do unto others as you would have them do unto you.
LUKE 6:31

The test of our progress is not whether we add more to the abundance of those who have much; it is whether we provide enough for those who have little.
FRANKLIN D. ROOSEVELT

As we saw in chapter 1, the concept of fairness is central to the income inequality debate. Indeed, it is at the very core of the stability and proper functioning of a democratic society and the free market. Therefore, before we address the more specific issue of fairness in income distribution, we need to examine it in the broader context of social justice, i.e., equality, liberty, and utility in a democratic society, exploring various leading theories of distributive justice that seek to specify what is meant by a just distribution of goods (Fleischacker 2004b; Roemer 1996).

## 2.1 Philosophical Perspectives: Mill, Rawls, Nozick, and Dworkin

The fairness issue is a particularly challenging one as the term *fairness* is often used quite broadly and can mean different notions in different contexts, as can be seen from the literature on this subject (e.g., Sen and Foster 1997; Fehr and Schmidt 1999; Bolton and Ockenfels 2000;

Kaplow and Shavell 2001; Thomson 2007; Akerlof and Shiller 2009). This is one of the reasons the term is often used in quotes, as in "fair." It is a tricky concept as what is morally fair can be different from what is fair in an economic sense. For instance, fairness based on moral principles might require us to recognize all human beings as equals, but does this imply that all employees should receive equal salaries irrespective of their contributions in an organization? Our sense of economic fairness would tend to dictate otherwise. Our intuitive sense of economic fairness suggests that one's reward should be commensurate with the value of one's contribution, i.e., more pay for greater contribution.

Economists have generally avoided treating the fairness issue as a mainstream topic, as reflected in the comment by Robert E. Lucas, Jr. (Lucas 2002), a Nobel Laureate in Economics: "Of the tendencies that are harmful to sound economics, the most seductive, and in my opinion the most poisonous, is to focus on questions of distribution.... The potential for improving the lives of poor people by finding different ways of distributing current production is nothing compared to the apparently limitless potential of increasing production." Lucas is, of course, right about the importance of growth and efficiency, but fairness also matters a great deal to people in real life, as noted in chapter 1 (we discuss this further in sections 5.9 and 5.10). However, most standard economics textbooks sidestep the fairness issue, focusing more on productivity and efficiency. But it is important to note that there are encouraging recent trends, as reflected by the literature on the ultimatum game and related topics on fairness (Güth 1995; Fehr and Schmidt 1999; Nowak et al. 2000).

While mainstream economics has generally avoided discussing equality and fairness, political philosophers have embraced it wholeheartedly, making it the central theme in their theories of a democratic society, as we see from the work of Rawls, Nozick, Dworkin, Sen, and others in the past four decades. These scholars have shed light on the roles of the free market and the state in socioeconomic systems with respect to equality, liberty, and fairness. Following up on the groundbreaking work of Rawls (1971) and Nozick (1974), there has been extensive literature on these topics (see, e.g., Sen and Foster 1997; Dworkin 2002; Tomasi 2012; Spatscheck 2012).

Our purpose here is not to review these pioneering contributions at any length. Our aim is to only highlight the key concepts so that when we present our framework, we can better articulate the similarities and differences between our theory and theirs. In particular, we will limit ourselves to the theories by Mill, Rawls (and Dworkin, to some extent), and Nozick, who are leading proponents of utilitarian, egalitarian, and libertarian models of fairness, respectively. We do not go into the various nuances of these theories, which have been extensively analyzed and critiqued over the years. For the uninitiated, the *Stanford Encyclopedia of Philosophy* is an excellent place to start to delve more deeply into this area (van Eijck and Visser 2012; see also Lebacqz 1986). We will limit ourselves to the essential ideas in order to set the stage for our system-theoretic perspective.

While many great intellectuals, such as Jeremy Bentham, had advanced utilitarianism, John Stuart Mill's exposition defined the intellectual framework for discussing theories about utility, justice, and fairness. The essential idea of utilitarianism is that the "right" thing to do is that which produces the greater good or "greater happiness" for those concerned (Mill 1957, 10). However, this view was always at odds with the traditional notions of justice at an individual level, since the strictly utilitarian principle leads to situations where individual rights may be sacrificed for the collective good of the society. But to utilitarians there can be no theory of justice that is devoid of utility. Indeed, Mill considers justice to be that which promotes our well-being, "the most vital of all interests," and "therefore of more absolute obligation, than any other rules for the guidance of life" (Mill 1957, 67). Hence, there can be no discussion of justice without acknowledging the importance of the utility of preserving and promoting well-being. Thus, in the eternal tussle between utility and justice, Mill's utilitarian view favors utility over justice, thereby valuing the society's interests more than an individual's.

Stressing the importance of individuals and their liberties, rather than society as a whole as the utilitarians prefer to, with a particular concern for the weaker ones in a society, John Rawls proposed his theory in *Justice as Fairness* (Rawls 1971, 2001) as a liberal egalitarian alternative, expressed in two main principles of liberty and equality. According to the first principle, every individual is to have equal basic liberties. Rawls arrives at this fundamental principle of equality among people through the application of his "veil of ignorance" concept to a group of rational agents in the "original position."[1] In conformity with the social contract tradition, Rawls assumes that the individuals in the original position are equal, free, and independent.

The second principle has two components and is stated, in its revised re-statement (Rawls 2001, 42), as follows: "Social and economic inequalities are to satisfy two conditions: first, they are to be attached to offices and positions open to all under conditions of *fair equality of opportunity*; and second, they are to be to the greatest benefit of the least-advantaged members of society (the *difference principle*)."

Rawls organizes these principles in *lexical ordering*, prioritizing them as the *liberty principle*, *fair equality of opportunity*, and the *difference principle*. This order determines the priorities of the principles should they conflict in practice.

Thus, Rawls stipulates that a just society is one wherein the social and economic inequalities are to be arranged such that they are of greatest benefit to the least-advantaged members of society. He proposes the *maximin* procedure for the distribution, i.e., maximizing the minimum prospects of individuals in a given society. This form of distributive justice is to be enforced by some *strong central authority*, such as the state. Rawls does *not* inform us what the resultant income distribution would be in his egalitarian society, in quantitative terms, beyond stipulating the maximin procedure for the distribution.

Disagreeing with the premise, Rawls's Harvard colleague and libertarian, Robert Nozick (1974), rejects the role of a strong central authority, countering with a theory that argues that only a minimal state is justifiable. Like Rawls, Nozick stresses the importance of individual liberties. However, as opposed to Rawls's strong state, he argues that a minimal state will emerge naturally, guided by an invisible hand, and will function solely to protect individuals' basic rights. This is essentially a free-market view of a society (Friedrich von Hayek [1944] had defended a similar view, not so much from the perspective of fairness and justice, but out of concern for "the danger of tyranny that inevitably results from government control of economic decision-making through central planning." [Eberling, 1999, 31]). Both Rawls and the utilitarians argue in support of a strong state to ensure that goods are distributed justly either to protect the least advantaged (the Rawlsian view) or to ensure the greatest overall good (the utilitarian view).

However, Nozick argues that only a minimal state is justified whose role is to protect and preserve basic rights of individuals such as life, liberty, and property, and to maintain social order. He asserts that both the Rawlsian and the utilitarian notions of justice are in fact unjust, and that only the distribution, whatever it may be, that *naturally* arises from the numerous individual exchanges and decisions through a free-market mechanism is justified. One of the key roles of the minimal state is to ensure the proper functioning of such a free market. Like Rawls, Nozick also does *not* inform us what the resultant income distribution would be in his libertarian society, in quantitative terms, beyond claiming that whatever distribution the free market delivers is fair.

Another important egalitarian theory of distributive justice comes from Ronald Dworkin, who differs from fellow egalitarian Rawls in his emphasis on equal access to resources (Dworkin 1981a, 1981b, 2002; Rogers 2000; Waldron 2001). Dworkin's theory is often identified as one of the earliest in the luck egalitarian literature, though he himself called his theory resource egalitarianism.[2] Like Rawls's theory, Dworkin's theory of equality has two parts: first a general theory of justice and then an account of the principles of distribution. His theory of justice values equality over liberty, arguing that it is the sovereign virtue (Dworkin 2002). According to him, the ideal of liberty already has built into it the notion of respecting the freedom of others, i.e., equal freedom for all. Thus, the notion of liberty is founded on equality. According to Dworkin, one should show equal concern for each person's life and equal respect for the basis on which he has chosen to live it.

Regarding distributive justice, Dworkin follows Rawls to some extent but differs in important ways. Egalitarians have tended to be vague about what they consider should be distributed equally, but they generally suggest that it should be some measure of well-being, such as utility or happiness. In two powerful essays (Dworkin 1981a, 1981b), Dworkin disagrees, arguing for *equality of resources* over *equality of welfare*, with important individualistic qualifications. Differing from Rawls, he argues that distributive justice should discriminate between consequences resulting from individual choices, on one hand, and luck, on the other. Dworkin uses the term *ambitions* to represent the realm of our choices and what results from our choices (such as the choice to work hard or not, or to spend money on luxuries). His term *endowments* refers to the results of brute luck, or those things over which we have no control (such as genetic inheritance or an accident making someone disabled).

On the distribution of resources, like Rawls, Dworkin conducts a thought experiment, asking us to imagine a desert island community in which everyone, starting with the same purchasing power, is able to bid in auction for a share of the island's resources to arrive at a fair distribution. Dworkin proposes that people begin with equal resources but be allowed to end up with unequal economic benefits as a result of their own choices. He argues that people who choose to work hard to earn greater income should not be required to subsidize those choosing greater leisure and hence less income.

More recently, John Tomasi (2012) proposed in *Free Market Fairness* a kind of compromise between Rawlsian egalitarianism and Nozickian libertarianism, in which he argues that capitalistic economic freedom should be treated as a basic liberty. He has tried to retain certain appealing features from both theories, but this has led to some inconsistencies, according to his critics (Koppelman 2012; O'Neill and Williamson 2012).

It is important to recognize at this juncture that while all these scholars made groundbreaking contributions in laying the philosophical foundations of a fair society, none of their frameworks could answer the four fundamental questions we raised in chapter 1. Toward this goal, we begin the formulation of our theory from the next section onward.

## 2.2 A System-Theoretic Perspective

In essence, the theories of Mill, Rawls, Nozick, Dworkin, and others address the foundational and organizational principles that a well-ordered, well-functioning, fair democratic society must adopt to meet its societal goals. In *system-theoretic* parlance, these are the *design* and *control principles* of a socioeconomic system.

Since a society is a complex construct that stands for many things, it can be analyzed philosophically, sociologically, economically, and even theologically. In addition, it is a *dynamical system* with a very large number of citizens (often in the millions), who interact with one another as well as with the environment in myriad ways. In this regard, it also exhibits behaviors and characteristics of a *statistical, dynamical system* that cannot, and should not, be ignored. While a society of human beings is obviously very different from an engineering system of inanimate elements, it nevertheless shares certain common characteristics, at an appropriate level of abstraction, that may be profitably analyzed using concepts and techniques from systems engineering, game theory, statistical mechanics, and information theory. Therefore, it is reasonable to model a human society from these perspectives.

Mill, Rawls, Nozick, Dworkin, and others framed what constitutes a fair society as a philosophical or sociological problem, or one of political economy. While these philosophical and sociological approaches have made great contributions to further understanding of societal organization and functioning, not addressing this other essential nature of society, i.e., as a statistical dynamical system of interacting agents, has resulted in an incomplete understanding that has missed critical insights, as our theory demonstrates.

We address this deficiency by developing a system-theoretic framework, which we have named *statistical teleodynamics*, that integrates foundational concepts and techniques from game theory, statistical mechanics, information theory, and systems engineering with those in philosophy, sociology, and economics to yield a quantitative theory that reveals surprising and useful insights as well as testable predictions that can be verified with empirical data. Our theory's name comes from the word *telos*, which means "goal" in Greek. Since our agents, such as employees and corporations, are driven by their goals (to maximize utility or profit) in their dynamics of switching jobs or employees, we call such *goal-driven* systems *teleodynamical* systems. Because of the stochastic or statistical nature of dynamics, it is called *statistical teleodynamics*. This framework may be seen as a generalization of the concepts and laws of statistical thermodynamics, developed originally for systems of inanimate entities, such as molecules, about 150 years ago, to teleological systems, such as economic and societal systems.

I have described the different elements of this theory in four earlier papers (Venkatasubramanian 2007, 2009, 2010; Venkatasubramanian et al. 2015). Here we develop this theory further to a more complete form and compare it with the theories of Mill, Rawls, Nozick, and Dworkin, pointing out the similarities and differences.

### 2.2.1 A New Theory of Hybrid Society

It is important to recognize that there has never been, and never will be, a *purely* utilitarian society (à la Mill), egalitarian society (à la Rawls or Dworkin), or libertarian society (à la Nozick) in the real world. These are, of course, idealized model societies, which we don't expect to see in practice. We don't see them because, in their purest form, they are impossible to implement in a democracy. Let us, for instance, consider Rawlsian egalitarianism versus Nozickian libertarianism as the path to follow. By the very nature of a democratic society, there will be a plurality of opinions about which path to follow as both have their own advantages and disadvantages. As a result, it will be difficult to implement one system or the other in its entirety without any compromise.

So it is no surprise that we end up with a *hybrid society* that has elements from different political philosophies. This is just a matter of *survival of the fittest* —the pure versions simply can't survive in a democratic environment of empowered citizens with diverse opinions who demand their views be accommodated. As a result, a compromise solution emerges—the hybrid society—as the fittest candidate in this environment. We will have more to say about this later when we discuss our theory of teleological systems in chapters 5 and 8.

Most, if not all, modern free market–based democratic societies in real life, particularly Western democracies, are such hybrid societies. They typically incorporate Nozick-like free-market principles, suitably modified by a Rawlsian difference principle, e.g., tax and transfer policies, while making utilitarian cost-benefit trade-offs. Nevertheless, these foundational principles, in their pure form, are still very useful because they provide us, at the very least, moral guidance for the sensible choices that each society could make, offering us a menu of choices to select and combine.

It is also important to keep in mind that these formal theories of Rawls, Nozick, and others appeared only in the 1960s, 1970s, and later, but their central principles were already integral to these hybrid societies, and arguably had been so for centuries. For example, life, liberty, and the pursuit of happiness were enshrined in the U.S. Declaration of Independence in 1776 as inalienable rights. Whether formalized in a theory or not, people have generally had an intuitive understanding of, and a desire for, liberty, equality, fairness, and happiness, and they incorporated them in their practices, to varying degrees, as they formed societies. People's initial positions were further refined, iteratively through practical experience over several decades, in response to economic booms and busts, to social unrests demanding equality on a number of issues. Consider, e.g., the different amendments made to the U.S. Constitution (e.g., the Thirteenth Amendment abolishing slavery or the Nineteenth Amendment granting women's suffrage) as a result of such iterative improvements, or the new laws passed (e.g., Glass-Steagall) and institutions created (e.g., the Federal Reserve) to respond to economic crises.

So human societies were not waiting for the philosophers, sociologists, and economists to figure all this out formally first and then implement it in practice. The experimentation and iterative self-adjustment started from the earliest days of civilization. The intellectuals only caught up with the reality with their formal theories much later, as is often true in many scientific endeavors. Therefore, such hybrid societies have long been a reality, at least in the Western world.

Hence, there is a clear need for theories that deal with this reality. Among such theories, the works of Schmidtz (2006), Wright (2010), Gaus (2010), and Tomasi (2012) stand out. What we propose in this book is not as philosophically deep or rigorous as their theories or as those of Rawls, Nozick, Dworkin, Sen, and Nussbaum. But, as opposed to all these beautiful philosophical theories, which are qualitative in nature and hence don't yield quantitative predictions, we take a much more pragmatic perspective and seek to develop a simple, but quantitative, framework that makes testable predictions which can be verified using empirical data. Our goal is to identify fundamental principles and develop simple models that offer an appropriate coarse-grained description of a hybrid society and make predictions not restricted by society-specific details and nuances.

Motivated by such a pragmatic outlook, one might say an engineering perspective, we consider a hybrid utopia—an ideal version of such a hybrid society—and explore its distributive justice property from an income distribution perspective, i.e., how fairly this society distributes wage income. We don't offer any particular philosophical defense or justification for our hybrid society, along the lines of Rawls, Nozick, or Dworkin, other than to defend it as a pragmatic solution, commonly arrived at by citizens in real life, to a challenging decision-making problem. (That said, we do provide a system-theoretic justification, founded on the *principle of maximum fairness*, in chapter 8.) The various elements of this compromise solution (such as the equality and liberty principles) have been defended eloquently by its original proponents and require no further defense. The only aspect that requires justification is the reason for this "mixing and matching," which we justify on a pragmatic basis. So one might call our approach *pragmatic egalitarianism* or *pragmatic libertarianism*. The typical hybrid society in real life has different elements adopted from utilitarian, egalitarian, and/or libertarian societal models, combining them to varying degrees. Some societies are more egalitarian, as in the Scandinavian countries. Others are less so, such as in the United States. Most others are typically somewhere on this spectrum. So, in practice, what really is left for us as a choice is the extent of this compromise—how to balance the needs of the society and those of the individual.

We now outline the essential features of this hybrid society, and then develop an analytical model and analyze its behavior and predictions. First, this hybrid society is a free market–based society in the Nozickian sense, with utility-maximizing agents and profit-maximizing corporations. However, we modify it by applying a Rawlsian treatment and introduce a *minimum* wage as a way of protecting the working poor. Such a floor is widely adopted in Western democracies, including the United States. While this may not be as generous to the weak as the maximin strategy advocated by Rawls, it's a nod to the spirit behind the principle. Most Western democracies, however, do take a more progressive approach by implementing tax and transfer policies that benefit the less fortunate, but in our current model we do *not* consider the effect of taxes and transfers (this is a natural next step which we discuss in chapter 8). Then we infuse this socioeconomic system with an element from Smith's (and Mill's) utilitarian framework; we assume the agents—both individuals and corporations—strive to maximize their self-interests.

In the model hybrid society, we assume, as Rawls and Nozick do, that all citizens are equal, free, independent, and fully capable. They do not suffer from any disability, physical or mental. Thus, we do *not* address the important questions that arise in a society in which some citizens are disabled, one way or the other, requiring special treatment. These issues have been addressed by Amartya Sen, Martha Nussbaum, and others. Our theory, in its present form, is focused on able individuals who generally constitute the bulk of a population. Again, this is another natural extension discussed in chapter 8.

## 2.3  What Is the Purpose of a Society?

Let us now develop the theory of this ideal hybrid society from a systems engineering perspective. In a sense, this is perhaps the ultimate systems engineering challenge.

How do you engineer a society of humans? That is, how do you "design," "control," and "optimize" a human society?

At first glance, this seems preposterous—designing and controlling a society of humans as if they were mechanical or electrical parts in a machine. This also sounds downright scary, raising all kinds of dystopian alarms. But this is what the Founding Fathers were attempting to do in their American experiment, as they laid down the principles of a democratic society they wanted to live in, and leave for their descendants, in the Declaration of Independence and the Constitution. So, this is not that crazy or scary, after all!

We now approach this challenge—how to design (i.e., organize), control (i.e., govern), and optimize (i.e., maximize benefits under constraints) a society?—as if it were a proper systems engineering problem.

When engineers design a system, they start with a series of questions that address various aspects of the design and operation of the system. The first question is: What is (are) the purpose(s) of this system?

A system, which is an organized structure of entities (animate or inanimate), typically exists to accomplish some goal(s) or purpose(s). For example, a chemical plant is an organized structure of equipment with a goal to safely and efficiently manufacture some chemical needed by society. A corporation is a system of people (i.e., animate entities or agents), organized to efficiently perform their tasks and generate profit, which is its economic objective. The corporation's main purpose is to make money for its investors and shareholders.

A simpler example of an engineering system would be a pressure vessel, which is a collection of gas molecules (i.e., inanimate entities) in a sealed container. This is a thermodynamic system, obeying the laws of thermodynamics. Thus the molecules are not organized or structured in any particular manner inside the vessel, beyond the constraint that they must be inside the vessel, which is the extent of organization needed for the purpose here. The objective of this system is to build pressure that can later be used to perform some useful task.

A more complicated example is an entire refinery (of which the pressure vessel could be one part) whose purpose is to convert crude oil into valuable products (such as gasoline) and thus generate profit for the corporation that owns the refinery. This is also a thermodynamic system, but a more complex one.

Thus, systems, whether human-engineered, such as corporations or chemical plants, or naturally evolved, such as life-forms or societies, typically exist to accomplish some objective(s).

In addition, systems engineers would subsequently ask many questions such as (not necessarily in this order): (1) What are all the inputs and outputs? (2) What are the properties of the individual agents?

(3) How do the agents interact with one another? (4) How should they be structured? (5) How does one define the state of the system at any given time? (6) What are the key state variables and parameters? (7) How are they measured? (8) What are the conservation laws and constitutive relations, if any? (9) How would the state of the system evolve over time? (10) What are the variables to be controlled? (11) What is the system's environment? (12) How does the system interact with its environment? (13) Will the system settle into an equilibrium state over time? (14) What criterion determines equilibrium? (15) Is the equilibrium stable to disturbances? (16) What are the risks to the system's stability? (17) What are the efficiency-robustness and cost-benefit trade-offs in various design and operational choices?

While there are major differences between a society (or an economic system) and an engineering system, several of these questions are still quite relevant in the design or organization of a human society or an economic system. Hence, it is useful to explore this avenue of inquiry.

Thus, from a systems engineering perspective, the first question to answer is: Given a society of human agents, who freely and dynamically interact among themselves and with the outside environment, what is (are) the purpose(s) of a society?

While this central question has been debated for centuries by philosophers, theologians, and sociologists, neither Rawls nor Nozick explicitly answers it in his theory. It can be seen to be implied in their discussions, but it is not stated as a clear objective as are the liberty and equality principles of Rawls. One infers that the purpose is well-being or happiness, whatever that may be. Liberty and equality per se cannot be the ultimate objectives of a society, for they naturally lead to the following question: Once every agent has liberty and equality, then what? What would the agents do with their liberty and equality? Thus, these are just means to an end. But what is that end?

Utilitarians like John Stuart Mill (1957) and Jeremy Bentham (1982), as well as other philosophers, notably John Locke (1689) and Samuel Johnson (1787), had suggested that the goal of life is the pursuit of happiness. This is also, by the way, the goal in Eastern philosophies, but it can get more complicated because of the spiritual nature of the discussion (Maharshi 2000; Bodhi 2005; DeLuca and Vivekananda 2006; Adidevananda 2009). While this goal is perhaps implied in the theories of Rawls and Nozick, it is not explicitly mentioned as the stated objective of a society. However, this is generally understood to be so as

C. S. Lewis (Lewis, 1996, 171) observes: "The State exists simply to promote and to protect the ordinary happiness of human beings in this life." This objective, of course, is the foundational principle of American society—the pursuit of happiness, empowered by life and liberty, the other two key principles of the tripartite motto.

As Hamilton (2008) points out, this Jeffersonian ideal has its origins in the writings of John Locke, and its philosophical lineage can be traced from Socrates, Plato, and Aristotle through the Stoics, Skeptics, and Epicureans. To quote Hamilton: "It appears not in the *Two Treatises on Government* but in the 1690 essay *Concerning Human Understanding*. There, in a long and thorny passage, Locke wrote:

> The necessity of pursuing happiness [is] the foundation of liberty. As therefore the highest perfection of intellectual nature lies in a careful and constant pursuit of true and solid happiness so the care of ourselves, that we mistake not imaginary for real happiness, is the necessary foundation of our liberty. The stronger ties we have to an unalterable pursuit of happiness in general, which is our greatest good, and which, as such, our desires always follow, the more are we free from any necessary determination of our will to any particular action, and from a necessary compliance with our desire, set upon any particular, and then appearing preferable good, till we have duly examined whether it has a tendency to, or be inconsistent with, our real happiness: and therefore, till we are as much informed upon this inquiry as the weight of the matter, and the nature of the case demands, we are, by the necessity of preferring and pursuing true happiness as our greatest good, obliged to suspend the satisfaction of our desires in particular cases."

We won't be addressing the voluminous literature on various interpretations and discussions of the tripartite motto here.[3] For our purposes, it means that one has the right to live a life, free to decide for oneself how to achieve happiness. We will take the widely held interpretation that this means the following: Objective 1-Life: security of life and property, including certain inalienable rights. Objective 2-Liberty: the freedom to live that life the way one chooses. Objective 3-Pursuit of happiness: to freely pursue happiness for oneself.

Objective 1 guarantees certain fundamental rights for the agents for their survival and existence. Objective 2 guarantees freedom of thought, decision, and action. Objective 3, in some sense, is the main objective of the society—objectives 1 and 2 are enablers to realize objective 3. In the American model, society allows an individual to pursue happiness, but with no guarantees on the outcome. Thus, different people may end up with different levels of happiness, or with none at all, in their lives. One is only provided with an opportunity to pursue it.

From a systems engineering perspective, reflecting on the list of questions cited above, objective 1 essentially defines the static properties of the agents, while objective 2 defines the dynamic aspects such as decisions, actions, and interactions that affect the temporal evolution of the agents and the system. In some sense, objective 1 defines the present while objective 2 refers to mechanisms for the evolution of the present state to a future state, in purely qualitative terms. Objective 3 is the overall goal of the society.

## 2.4  BhuVai: A Hybrid Utopian Society of Universal Happiness

Let us now consider an ideal, utopian, hybrid society, which we call BhuVai, inspired by a concept in Hindu philosophy of a paradise on Earth, called *Bhuloka Vaikuntham*. In this society, all are treated as equals. Everyone enjoys the same rights associated with life and liberty, similar to the individuals in Rawlsian and Nozickian societies. In addition, BhuVai strives to ensure that everyone also enjoys the same level of happiness. With this ambitious objective, BhuVai differs dramatically from the Rawlsian, Nozickian, or other societies that have been proposed before.

This objective is, of course, not achievable in real-life societies. While this may seem to be an impractical, even laughably silly, objective of little utility, it turns out to be quite useful as a reference state, as we shall demonstrate.

The logic for this objective is as follows. In our utopia, everyone is making a valuable contribution to the overall functioning of the society. The sanitation workers, teachers, accountants, farmers, doctors, chefs, bankers, lawyers, executives, police and military personnel, cobblers, poets, painters, actors, etc., are all making a valuable contribution to the successful functioning of the society. Imagine what life would be like if sanitation workers stopped clearing garbage, if farmers stopped growing food, if teachers refused to teach, and so on. Any society would come to a grinding halt. If the bottom 90% refused to cooperate, did not perform their jobs effectively, or did not perform them at all, then the top 10% would find life to be very difficult, no matter how much money they have. In this sense, everyone is making a valuable contribution to the overall functioning of a society.

Therefore, in our utopia, the goal is to make all people *equally happy* for performing their jobs well and making their contributions. Wouldn't it be wonderful to have a society where all the people pursue their lives the way they wish, making a valuable contribution and enjoying the same level of happiness as everyone else in the society? This idea runs along the lines of Adam Smith's sentiments (Smith 1976, 664): "allowing every man to pursue his own interest his own way, upon the liberal plan of equality, liberty, and justice." This is an *envy-free* society. Dworkin discusses the envy test for his *equality of resources* proposal (Dworkin 1981a, b). Here we have it in the context of *equality of welfare*. No one in this utopia would want to trade her life for someone else's, since she is already as happy as everyone else is, leading the life she wants to lead. What more could we ask of a society?

Past work has generally shied away from the objective of equality of welfare, well-being, or happiness for the simple reason that it is very hard to measure. Happiness is a fuzzy and elusive state of an agent, depending on a whole array of complex factors. So no one could measure it, correctly and consistently, across a population. That's why either this objective is generally avoided, as in the theories of Rawls and Nozick, or the focus shifts to equality of resources, as Dworkin proposed. Even the aforementioned giants of philosophy and statecraft limited themselves to just the pursuit of happiness as a fundamental right. They did not consider the realization of happiness as a fundamental right, let alone everyone achieving the same amount of it! It is just plain silly, a fool's errand, to propose this as a societal objective. Thus, one can raise a number of objections to this idea.

But even Dworkin concedes that this is a laudable objective. As he observes (Dworkin 1981a):

> It does not follow, however, that the ideal of equality of welfare, on any interpretation, is either incoherent or useless. For that ideal states the political principle that, so far as is possible, no one should have less welfare than anyone else. If that principle is sound, then the ideal

of equality of welfare may sensibly leave open the practical problem of how decisions should be made when the comparison of welfare makes sense but its result is unclear. It may also sensibly concede that there will be several cases in which the comparison is even theoretically pointless. Provided these cases are not too numerous, the ideal remains both practically and theoretically important.

While measuring happiness, or some such subjective well-being, has generally been considered impractical and suspect, recent advances make it indeed a reasonable exercise to perform. For instance, under the auspices of the United Nations, the first *World Happiness Report* was published in support of the 2012 United Nations High Level Meeting on Happiness and Well-Being. We quote from the most recent report, published in 2015: "The world has come a long way since the first World Happiness Report launched in 2012. Increasingly happiness is considered a proper measure of social progress and goal of public policy. A rapidly increasing number of national and local governments are using happiness data and research in their search for policies that could enable people to live better lives. Governments are measuring subjective well-being, and using well-being research as a guide to the design of public spaces and the delivery of public services."

However, the objection that this objective is problematic in the real world is a valid one. That's why it's a utopian dream! But contrary to one's typical gut reaction that pursuing this line of inquiry is therefore futile, this impractical target is surprisingly useful, as we demonstrate, in engineering real-world societies by serving as a useful reference state in the following sense. Most people, perhaps even all, desire happiness more than anything else in their lives. Recognizing "happiness and well-being as universal goals and aspirations in the lives of human beings around the world," the U.N. General Assembly declared March 20 as World Happiness Day in 2012. As noted, it is even enshrined in the U.S. Declaration of Independence.

Therefore, a society where everyone is *equally happy* is the *best* a society can deliver to its citizens. It has achieved its purpose. So this is the target to shoot for. We can't do any better than this. This is the utopia everyone would love to live in. So, if a real-world society comes close to this utopia, it is doing very well for its citizens. This is a key insight, which we shall exploit in our theory.

But how can we know whether a society has accomplished this? How do we measure how close a given society is to this ideal target? Can we even measure this?

Surprisingly, we can, in the context of distributive justice. This is what we will accomplish in the rest of the book. As noted, happiness depends on a whole array of complex factors. It clearly depends on one's health (mental, physical, emotional, and perhaps even spiritual), income and wealth, family and friends, work environment, etc. In our theory, we limit ourselves to only the economic aspect. We are not accounting for the other factors even though they are obviously very important. Our focus is entirely on the happiness derived from work in the form of economic rewards.

The rest of the book is focused on the question of *distributive economic justice* in this model hybrid society, particularly how wage income would be distributed in a fair manner. Therefore, we now proceed to develop a quantitative, analytical model of our ideal hybrid society, BhuVai, and analyze its distributive justice property with respect to wage income distribution in chapter 3.

# CHAPTER THREE

# Distributive Justice in a Hybrid Utopia

A perfect formulation of a problem is already half its solution.
DAVID HILBERT

---

All models are wrong, but some are useful.
GEORGE E. P. BOX

In this chapter, we begin the development of the mathematical formalism that answers the four fundamental questions we raised in chapter 1. As noted, none of the mainstream economic theories or the econophysics models or the philosophical frameworks answer these questions.

The goal of our model society, a hybrid utopia called BhuVai, is to empower all citizens to pursue their visions of happiness and to realize those visions in practice. The happiness thus enjoyed by everyone is at the same level. While overall happiness certainly depends on a number of factors such as physical, emotional, and economic health, we restrict ourselves to economic health as our focus is on distributive economic justice. We assume that the citizens in BhuVai are physically and emotionally healthy and therefore are happy as far as these factors are concerned. Our focus is entirely on a person's work—what it contributes to the society and the happiness derived from it in the form of economic rewards. Since our focus is on distributive economic justice, particularly how income would be distributed in a just manner in our model society, we assume that all other requirements of social justice have already been satisfied in BhuVai.

## 3.1 Income Distribution in BhuVai: Formulating the Problem

How, then, would distributive justice be realized in BhuVai? As noted, since different people have different skills and therefore make different contributions to a society, we expect people to be compensated unequally monetarily. So a certain level of income inequality is to be expected. But what is the fairest level of income inequality?

Recall that our hybrid utopia is part Rawlsian and part Nozickian. Following the Nozickian feature, we let this income distribution be achieved through the free-market mechanism. As Nozick argues, whatever income distribution that *naturally arises* as a result is the fairest outcome according to his libertarian justifications. As noted, the only modification we make to his purely free-market view is to introduce a Rawls-like condition that there is a *minimum-wage* floor to protect the working poor. Most free-market democracies exhibit these two characteristics. Therefore, the distributive justice question is: In our ideal hybrid free-market environment, what level of inequality should arise naturally? This is the question we address in this chapter and chapter 5.

Note that in our utopia, different levels of incomes *do not* result in different levels of happiness. For instance, even though an artist may make less than, say, a hedge fund manager, they are both equally happy. This is the key feature of our hybrid utopia (this feature is explored in greater detail in section 8.3 after we have introduced and discussed our framework).

### 3.1.1 Ideal Versus Real-Life Free Market

To determine this income distribution, we propose the following gedankenexperiment, in which we study a competitive, dynamic free-market environment in BhuVai, comprising a large number of utility-maximizing rational agents as employees and profit-maximizing rational agents as corporations. Let us assume an ideal environment in which the free market is perfectly competitive, transaction costs are negligible, and no externalities are present. There are no biases due to race, gender, religion, etc. We assume that no single agent (or small group of agents), whether an employee or a company, can significantly affect the market dynamics; i.e., there is no rent seeking or market power. In other words, our ideal free market is a level playing field for all its participants.

We also assume that neither the companies nor the employees engage in illegal practices such as fraud, collusion, and so on. We also assume moderate scarcity of resources. We do not consider extreme situations in demand or in supply. We do not consider cases wherein there is an extremely high demand, or extremely low demand, for any skills. Similarly, we do not consider cases wherein there is an abundant supply, or an extreme scarcity, of skills. We consider such situations as exceptions rather than the norm in a typical market. We do not consider the effect of taxes as we compare our predictions with real-world pretax income distribution data from Piketty and coworkers in chapter 6. (We plan to address the effect of taxes and transfers in future work, which we discuss in chapter 8.)

In our ideal free market, employees are free to switch jobs and move between companies in search of better utilities. Similarly, companies are free to fire and hire employees to maximize their profits. We also assume that a company needs to retain all its employees in order to survive in this competitive market environment. Thus, a company will take whatever steps necessary, allowed by its constraints, to retain its employees. Similarly, all employees need a salary to survive, and they will do whatever is necessary, allowed by certain norms, to stay employed.

It is important to emphasize that the free market itself, ideal or otherwise, is a human creation and "does not exist in the wilds beyond the reach of civilization" (Reich 2015, 4). As Reich (2015, 3–5) observes, "Few ideas have more profoundly poisoned the minds of more people than the notion of a 'free market' existing somewhere in the universe, into which government 'intrudes.' ... A market—any market—requires that the government make and enforce the rules of the game. In most modern democracies, such rules emanate from legislatures, administrative agencies, and courts. Government doesn't 'intrude' on the 'free market.' It creates the market." In our utopia, the government has defined the rules of the game in such a way that the resulting free market behaves ideally.

Real-life free markets are, of course, not ideal. Biases of various kinds are common. As Bebchuk and Fried (2006), Reich (2015), and Stiglitz (2015) argue, rent-seeking behavior by executives has contributed to excessive CEO compensation in recent years, while the deteriorating bargaining power of rank and file employees due to weakened unions has resulted in their diminished share of the income pie. Nevertheless, it is useful to analyze the behavior of an ideal free market as it provides a reference state, identifying what the outcome should be under ideal conditions. We can then compare this result with the outcomes seen in real life, and both assess and correct nonideal behavior through appropriate policy prescriptions. Simply put, we need to understand how an ideal free market behaves before we can analyze all the complexities of a real-life free market.

### 3.1.2  A Gedankenexperiment *in the Ideal Free Market*

In this ideal free market, consider a company, named iAvatars Inc., with $N$ employees and a payroll of $M$, with an average salary of $S_{ave} = M/N$. We assume that there are $n$ categories of employees, ranging from low-level employees to the CEO, contributing in different ways to the company's overall success and value creation. All the employees contribute to the company's overall success in their own ways. The cleaning crew keeps the premises neat, the secretarial staff helps with organization and communication, engineers develop products and services, marketing and sales personnel bring new orders, accounting and finance personnel mind the books, management focuses on a winning corporate strategy and execution, and so on. How do we value all employees' contributions and reward them fairly?

All employees in category $i$ contribute value $V_i$, $i \in \{1, 2, \ldots, n\}$ such that $V_1 < V_2 < \cdots < V_n$. Let the corresponding value at $S_{ave}$ be $V_{ave}$, occurring at category $s$. Since employees are contributing unequally, they need to be compensated differently, commensurate with their relative contributions to the overall value created by the company. Instead, iAvatars has an egalitarian policy that all employees are equal, and therefore it pays all of them the same salary $S_{ave}$ irrespective of their contributions. The salary of the CEO is the same as that of a janitor. This salary distribution is a sharp vertical line at $S_{ave}$, as seen in figure 3.1(a), which is the Kronecker delta function.

While this may seem fair in a social or moral justice sense (this distribution has a Gini coefficient of 0), clearly it is not in an economic sense. If iAvatars were the only company in the economic system, or if it were completely isolated from other companies in the economic environment, then the employees would be forced to continue to work under these conditions as there would be no other choice.

However, in an ideal free-market system there are choices. Therefore, employees who contribute more than the average, i.e., those in

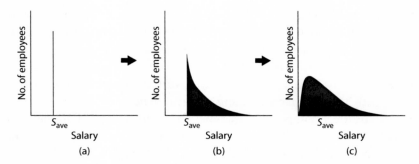

*Figure 3.1* Spreading of the pay distribution under competition in an ideal free-market environment. (a) Initial pay distribution, (b) Pay distribution spreading to the right under competition, and (c) Pay distribution after spreading to the right and left under competition.

value categories $V_i$ such that $V_i > V_{ave}$ (e.g., senior engineers, vice presidents, CEO), who feel that their contributions are not fairly valued and compensated by iAvatars will therefore be motivated to leave for other companies where they are offered higher salaries. Hence, to survive, iAvatars will be forced to match the salaries offered by others to retain these employees, thereby forcing the distribution to spread to the right of $S_{ave}$, as seen in figure 3.1(b).

At the same time, the generous compensation paid to all employees in categories $V_i$ such that $V_i < V_{ave}$ will motivate candidates with the relevant skill sets (e.g., low-level administration, sales, and marketing staff) from other companies to compete for these higher-paying positions in iAvatars. This competition will eventually drive the compensation down for these overpaid employees, forcing the distribution to spread to the left of $S_{ave}$, as seen in figure 3.1(c). Eventually, we will have a distribution that is not a delta function, but a broader one where different employees earn different salaries depending on the values of their contributions as determined by the free-market mechanism. The funds for the higher salaries now paid to the formerly underpaid employees (i.e., those who satisfy $V_i > V_{ave}$) come out of the savings resulting from the reduced salaries of the formerly overpaid group (i.e., those who satisfy $V_i < V_{ave}$), thereby conserving the total salary budget $M$.

Thus, we see that concerns about fairness in pay cause the natural emergence of a more equitable salary distribution in a free-market environment through its self-organizing, adaptive, and evolutionary dynamics. The point of this analysis is not to model the exact details of the free-market dynamics, but to show that the notion of fairness plays a central role in driving the emergence and spread of the salary (in general, utility) distribution through the free-market mechanism.

Even though an individual employee cares only about her utility and no one else's, the collective actions of all the employees, combined with the profit-maximizing survival actions of all the companies, in an ideal competitive free-market environment of supply and demand for talent under resource constraints, lead toward a more fair allocation of pay, guided by Adam Smith's "invisible hand" of self-organization. This is the invisible hand process that Nozick refers to in his libertarian proposal. So, instead of state-imposed maximin allocation aimed at achieving distributive justice, as Rawls recommends, the free-market dynamics determines the level of income inequality naturally. In chapter 5 we will prove that this is the fairest outcome not only in the Nozickian sense but also in the system-theoretic sense.

We have used salary as a proxy for utility (i.e., happiness derived from work) in this example to motivate the problem (we use the term *utility* to stand for happiness, welfare, or well-being in the rest of this book). In general, utility is a complicated aggregate that depends on a host of factors, some measurable, others not. Obviously, pay (i.e., total compensation including base salary, bonus, options, etc.) is an important component of utility, generally the dominant one. Other components include the quantity and quality of the effort, title and peer recognition, competition and job security, career and personal growth opportunities, retirement and health benefits, company culture and work environment, job location, and so on, not necessarily in that order.

Given this free-market dynamics scenario, three important questions arise: (1) Will the self-organizing free-market dynamics lead to an equilibrium salary distribution, or will the distribution continually evolve without ever settling down? (2) If there exists an equilibrium salary distribution, what is it? (3) Is this distribution fair? These are essentially the same as the first two questions from chapter 1, refined by our gedankenexperiment.

Our knowledge of the free-market dynamics, as well as of distributive justice, is incomplete, in a fundamental way, without an answer to these critical questions. This requires a theoretical understanding of the free-market dynamics, at a reasonable level of depth, particularly from the bottom up, agents-based, microeconomic perspective as described above. Given the obvious complexity of real-life free-market dynamics, it is unrealistic to expect to develop a theory, and the associated models, that will address all the details and nuances. Therefore, our goal is to develop an analytical framework that identifies the key concepts and general principles, that models free-market dynamics under ideal conditions, and that answers these central questions. We develop such a framework in the following sections.

## 3.2 A Microeconomic Game-Theoretic Framework: "Restless" Agents Model

We address these questions by developing a microeconomic framework based on potential game theory (There is a brief introduction to potential game theory below.) Continuing with the scenario described above, we assume that all employee agents get "dissatisfied," now and then, in their current positions. After working in the same position for a while, every employee feels that she has learned all that could be learned and so it is time for her to move up in her career. Every employee feels, sooner or later, that she could be and should be doing better, given her talents and experience, in her company or elsewhere. As a result, employees are on the lookout for job opportunities to improve their utilities. That is, these *utility-maximizing, fairness-seeking, teleological* agents become restless, now and then, itching to move to a better position.

Let us now examine the basic question of why people seek employment. At the most fundamental level, survival is to be able to pay bills now so that they can make a living, with the hope that the current job will lead to a better future. One hopes that the present job will lead to a better one, acquired based on the experience from the current job, and to a series of better jobs in the future, hence to a better life. This opportunity for a better future, with the expectation of *upward mobility*, is valuable to most, if not all, of us. This is the utility of having a *fair shot* at better future prospects. We are, of course, prepared to put in the requisite effort to earn such a life. This effort includes not only the effort we would put into our present jobs but also the effort (and time and money) we invested in the past to acquire the requisite education, skills, and experience. Thus, the utility derived from a job is made up of two components: the *immediate* benefits of making a living (i.e., "present utility") and the prospects of a better *future* life (i.e., "future utility").

Hence, we propose that the overall or effective utility from a job is determined by three dominant elements: (1) utility from salary, (2) disutility from effort, and (3) utility from a fair opportunity for recognition and career advancement. By effort, we do not mean just the physical effort alone, as we discuss later, even though that is a part of it. This overall utility is the microeconomic foundation on which we build our theory to predict and explain the emergent macroeconomic consequences in pay distribution in an ideal free-market environment.

Thus, the effective utility for an agent is given by

$$h_i(S_i, E_i, N_i) = u_i - v_i + w_i \qquad (3.1)$$

where $h_i$ is the effective utility (i.e., happiness from work) of an employee earning a salary $S_i$ by expending an effort $E_i$, while competing with $N_i - 1$ other agents in the same job category $i$ for a fair recognition of one's contributions. Here $u(\cdot)$ is the utility derived from salary, $v(\cdot)$ is the disutility from effort, and $w(\cdot)$ is the utility from a fair opportunity for a better future. Every agent tries to maximize effective utility by picking an appropriate job category $i$.

### 3.2.1 Utility of a Fair Opportunity for a Better Future

The first two elements are rather straightforward to appreciate, but the third requires further discussion. The first two elements model the tendency of an employee to maximize one's utility from salary while minimizing the effort put into receiving it.

As for the third element, consider the following scenario. A group of freshly minted law school graduates (totaling $N_i$) have just been hired by a prestigious law firm as associates. They have been told that one of them will be promoted to partner in eight years depending on their performances. Let us say that the partnership is worth $\$Q$. So any associate's chance of winning the coveted partnership goes as $1/N_i$, where $N_i$ is the number of associates in her peer group $i$, her *local competition*. Therefore, her expected value for the award is $Q/N_i$, and the utility derived from it goes as $\ln(Q/N_i)$ because of diminishing marginal utility. Therefore, the utility derived from a fair opportunity for recognition and career advancement in a competitive environment is[1]

$$w_i(N_i) = -\gamma \ln N_i \qquad (3.2)$$

where $\gamma$ is a parameter. This equation models the effect of *competitive interaction* among agents. In considering the society at large, this equation captures the notion that in a fair society, an appropriately qualified agent with the necessary education, experience, and skills should have a fair shot at growth opportunities irrespective of her race, color, gender, and other such factors; i.e., it is a *fair competitive* environment. This is the utility derived from *equality of access* for a better life, for *upward mobility*. This is the utility of having a fair shot at a better future. Category $i$ would correspond to her qualification category in the society. The other agents in that category are the ones she will be competing with for these growth opportunities. We note that this mathematical expression captures what Rawls mentioned qualitatively in the first part of his second principle, *fair equality of opportunity*. We weren't particularly looking to model this Rawlsian requirement, but, as we see, it emerges naturally in our formulation as an integral feature.

### 3.2.2 Modeling the Disutility of a Job: Net Utility's Inverted U-profile

For the utility derived from salary, we employ the commonly used logarithmic utility function

$$u_i(S_i) = \alpha \ln S_i \tag{3.3}$$

where $\alpha$ is a parameter. As for the second element, every job has a certain disutility associated with it. This disutility depends on a host of factors such as the investment in education needed to qualify oneself for the job, the experience to be acquired, working hours and schedule, quality of work, work environment, company culture, relocation anxieties, etc., which are often difficult, if not impossible, to quantify accurately in real life. While there is considerable prior work on modeling the disutility of effort when a metric for effort is available (Fehr et al. 1998; Laffont and Tirole 1993; Cadenillas et al. 2002; Böhringer and Löschel 2013; Dossani 2002), there isn't much when such a metric is absent.

In the absence of such a metric, typically one compensates for these different uncertain components of the disutility of a new job by negotiating a salary package that would make it worth the undertaking. Thus, in practice, one intuitively uses salary as a proxy to impute and estimate the "cost" of the job, thereby estimating the remuneration required to compensate for the disutility of effort. Note, again, that by effort we mean not just the hours put into performing the job, but all the prior investment in education and experience to qualify for the position as well as the other adjustments and sacrifices to be made in the new position. There is empirical evidence that supports this line of reasoning in the work of Stratton (2001) and Ahituv and Lerman (2007), who have demonstrated that effort correlates with $\ln(S)$.

Combining this with the commonly used quadratic disutility from effort (Laffont and Tirole 1993; Cadenillas et al. 2002; Böhringer and Löschel 2013; Dossani 2002; Holmstrom and Milgrom 1991; Nalebuff and Stiglitz 1983; Laffont and Tirole 1988, 1987; Zabojnik 1996), we propose the following form for the second element:

$$v_i(E_i) = \beta(\ln S_i)^2 \tag{3.4}$$

where $\beta$ is a parameter. Our formulation is also consistent with the conditions imposed on effort $E$ as a function of salary, $E(S)$. According to Katz (1986) and Akerlof and Yellen (1986), $E(S)$ should satisfy the following conditions:

$$\frac{dE}{dS} > 0 \quad E(0) \leq 0 \quad \text{and} \quad \frac{S}{E} \times \frac{dE}{dS} \text{ is decreasing} \tag{3.5}$$

Our effort function $E(S) = \ln S$ satisfies all three conditions:

1. $\dfrac{dE}{dS} = \dfrac{1}{S} > 0$

2. $E(0) = -\infty < 0$

3. $\dfrac{S}{E} \times \dfrac{dE}{dS} = \dfrac{S}{\ln S} \times \dfrac{1}{S} = \dfrac{1}{\ln S}$ is decreasing.

There is also another simpler, more intuitive way to arrive at equation (3.4). Combine $u$ and $v$ to compute

$$u_{\text{net}} = au - bv \quad a \text{ and } b \text{ are positive constant parameters}$$

which is the net utility (i.e., net benefit or gain) derived from a job after accounting for its cost. Typically, net utility will increase as $u$ increases (e.g., because of salary increase). However, generally, after a point, the cost has increased so much—due to personal sacrifices such as working overtime, missing quality time with family, giving up on hobbies, job

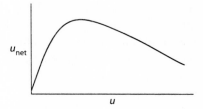

*Figure 3.2* Net utility: inverted-U curve.

stress resulting in poor mental and physical health, etc.—that $u_{net}$ begins to decrease after reaching a maximum (see figure 3.2).

The simplest model of this commonly occurring inverted-U profile is a quadratic function, as in

$$u_{net} = au - bu^2$$

Since $u \sim \ln(S)$, we get equation (3.4). We believe this is the simplest expression that captures this typical, widely observed behavior of net utility—increasing first and then decreasing. Therefore, our formulation for effort is a reasonable one supported by empirical evidence as well as by intuitive and theoretical expectations.

Note that most things in life in which the benefit gained has a cost associated with it have this inverted-U curve trend seen in figure 3.2. Consider, e.g., taking some medicine to cure an ailment. If you are suffering from a severe headache, taking one Tylenol might make you feel a little better, but taking two might help more. This doesn't mean that taking ten would help a lot more! It could actually make you sicker, triggering a whole set of other, more serious problems such as kidney disease and bleeding in the digestive tract. As the dosage increases beyond a critical point, the benefit begins to go down, sometimes dramatically. This is so because the cost of the treatment (in the form of negative side effects) begins to exceed the benefit.

Another such example can be seen in exercising. If running an hour per day is good for you, running ten hours per day, every day, is not necessarily great for you (i.e., for most people, excluding gifted long-distance athletes), because of the serious wear and tear on various body parts. The hedge fund titans who went bust in the subprime crisis learned this annoying little fact the hard way when they relied on excessive leverage to generate their outsized returns. Leverage of

2 to 3 (i.e., taking moderate risk) can be helpful to juice up the returns ordinarily, but when it is over 30, the downside is just disastrous. There are many such examples from all walks of life. In summary, more isn't always better, because most benefits usually have costs associated with them. Our utility function captures and models this essential and near-universal trait in most things in life.

### 3.2.3 Effective Utility Enjoyed from a Job

Combining all three, we have

$$h_i(S_i, E_i, N_i) = \alpha \ln S_i - \beta (\ln S_i)^2 - \gamma \ln N_i \tag{3.6}$$

where $\alpha, \beta, \gamma > 0$.

In general, $\alpha$, $\beta$, and $\gamma$, which model the relative importance an agent assigns to these three elements, can vary from agent to agent. However, we first examine the ideal situation in which all agents have the same preferences and hence treat these as constant parameters. (In sections 5.10 and 6.5, we relax this requirement and consider other cases. We also discuss what these parameters imply in greater detail in section 8.4.) As noted, presumably there are other expressions one could use to model these three elements, but the choices we made have interesting properties, revealing important insights and connections, as we shall see shortly.

To move to a job with better utility, an agent needs job offers. So the employee agents constantly gather information and scout the market, and their own companies, for job openings commensurate with their skill sets, experiences, and career and personal goals. Similarly, the company agents (e.g., through the human resources department) also conduct similar searches, looking for opportunities to fire and hire employees so that their profits may be improved.

At any given time, an employee agent is faced with one of five job options: (1) no new job offer is available, (2) the new offer has the *same* utility as the current one, (3) the new offer has *less* utility than the current one, (4) the new offer has *greater* utility, or (5) the employee agent is let go from the current job (i.e., *zero* utility). The agent's best strategies for the five options are as follows: for (1), (2) and (3), the agent stays put in the current position at the current utility; for (4) the agent accepts the new offer; and for (5) the agent leaves the company and looks for a new position. Each agent's strategy is independent of what other agents are doing.

We are now ready to answer these questions: (1) Is there an equilibrium salary distribution? (2) If yes, what is it?

## 3.3 Potential Game Formulation of Free-Market Dynamics

We answer these questions by using two different, but related, approaches. The first approach, which we call the *bottom-up* perspective, uses concepts and techniques from game theory. The second approach, which we call the *top-down* perspective, uses concepts and techniques from statistical mechanics and information theory. We develop the former in this chapter and the latter in chapter 5.

### 3.3.1 Game Theory: Players, Strategies, and Equilibrium

Before we jump into the technical details, let us start with an informal background to game-theoretic concepts for nonexperts.[2] The problem from section 3.1 is the following: Given a set of strategically interacting rational agents (called *players*), where each agent is trying to decide and execute the best possible course of action that maximizes the agent's outcome (called *payoff*) in light of similar strategies executed by the other agents, we are interested in predicting what strategies will be executed and what outcomes are likely. In other words, what are the possible outcomes of the strategic behaviors of the employees and the company in our thought experiment?

Game theory is a mathematical framework for answering this question by systematically analyzing and predicting the decision-making strategies, the dynamic behavior, and the likely payoffs of players, using models of conflict and cooperation. Just as *probability theory* originated from the study of pure gambling *without* strategic interaction, *game theory* originated from the formal study of *strategic games*, such as chess or Le Her, a card game. While some game-theoretic ideas could be traced as far back as the Babylonian Talmud (Walker 2012), most people generally credit the groundbreaking book *The Theory of Games and Economic Behavior*, by John von Neumann and Oskar Morgenstern, published in 1944, as the origin of the mathematical study of game theory. Great progress was made in the subsequent two decades, which is when famous game theory problems such as the prisoner's dilemma[3] were invented. This was also the period when John Nash made his seminal contribution (Nash 1950, 1951), which is central to the question we raised above.

Drawing the distinction between *cooperative* games, in which binding agreements can be made, and *noncooperative* games, where they are not feasible, Nash proved the existence of a *strategic equilibrium*, now called the *Nash equilibrium*, for *noncooperative* games. In a Nash equilibrium outcome, no player can unilaterally move to improve her own payoff. In other words, players have *no incentive* to change their actions, since their current strategy is the *best* they can do, given the actions of the other players.

The concept of an *equilibrium*, whose existence or nonexistence can be predicted through a systematic analysis of the game, is one of the most fundamental results in game theory with far-reaching implications in economics, sociology, political science, biology, and engineering. All these disciplines are riddled with situations in which one is interested determining the eventual outcome of a set of players making strategic decisions—one wants to know whether this ultimate outcome is an equilibrium state. This is exactly the same question we have in our problem above.

Interestingly, it is the same question that arises when one is dealing with a large number of entities that are interacting *not* strategically but *randomly*, such as molecules of a gas enclosed in a container. We want to know the equilibrium state for the gas and what criterion determines it. This problem is solved by using a mathematical framework called statistical mechanics, which is founded on probability theory. As noted, probability theory originated from the study of games of chance, whereas game theory originated from games of strategy. Since both mathematical frameworks define and determine an equilibrium state, it is natural to ask whether there is any connection between the two approaches. As it turns out, there is a deep, beautiful, and hitherto unknown connection with surprising and useful insights. We explore this in chapter 5.

Nash was recognized for his pioneering contributions with the Nobel Prize for Economic Sciences (officially, the Sveriges Riksbank Prize in Economic Sciences in Memory of Alfred Nobel), in 1994, which he shared with fellow game theorists John C. Harsanyi and Reinhard Selten. This was the first Nobel awarded for work in game theory. Some readers might recall that the life and achievements of John Nash were celebrated in the Hollywood movie *A Beautiful Mind*.

### 3.3.2  Population Games and the Potential Function

Among the many classes of games, there is a class called *population games* that is particularly relevant to the situation described in our gedankenexperiment.[4] This class studies games with a large number of players, as in our example. It provides an analytical framework for studying the strategic interactions of a population of agents with the following properties, as described by Sandholm (2010): (1) The number of agents is large. (2) Individual agents play a small role—any one agent's behavior has only a small effect on other agents' payoffs. (3) Agents interact anonymously—each agent's payoffs depend on only the opponents' behavior through the distribution of their choices. (4) The number of roles is finite—each agent is a member of one of a finite number of populations. (5) Payoffs are continuous—this property ensures that very small changes in aggregate behavior do not lead to large changes in payoffs. We note that these properties more than adequately account for the features of our game played in the gedankenexperiment.

In games like prisoner's dilemma, with a small number of players, the typical game-theoretic analysis proceeds by systematically evaluating all the options every player has, determining the payoffs, writing the payoff matrix, identifying the best responses of all players, and identifying the dominant strategies (if present) for everyone, and then finally reasoning whether there exists a Nash equilibrium (or multiple equilibria). However, for population games, given the large number of players, this procedure is not feasible for several reasons. For instance, a player may not know of all the strategies others are executing (or could execute), and their payoffs, in order for her to determine her best response.

Fortunately, as it turns out, for some population games one can identify a single scalar-valued global function, called a *potential*, that captures the necessary information about payoffs. The *gradient* of the potential is the *payoff* or the *utility*. Such games are called *potential games*. The origins of potential game theory can be traced to the concept of a *congestion game* proposed by Rosenthal (1973). Rosenthal proved that any congestion game is a potential game, and Monderer and Shapley later proved the converse (Monderer and

Shapley 1996)—for any potential game, there exists a congestion game with the same potential function.

A SIMPLE EXAMPLE: TRAFFIC CONGESTION A congestion game commonly arises in analyzing problems such as congestion in traffic networks. We illustrate this with a simple example. Consider driving from home to work, where you have two route choices. Route $R_1$ is wider with a width of $a_1$, and route $R_2$ is narrower with a width of $a_2$ (that is, $a_1 > a_2$). Assume that the travel time $t$, from home to work, goes *inversely* as the width; i.e., the wider route tends to be faster than the narrower route. Let us further assume that both $a_1$ and $a_2$ have some lower and upper bounds. The bounds don't really matter for the reasoning below; they are only there to avoid the unrealistic outcomes such as $t_1 = 0$ or $t_2 = 0$ when $a_1 = \infty$ or $a_2 = \infty$. Let us also assume that $t$ increases *linearly* with the amount of traffic on the route, i.e., the congestion. In other words, $t$ increases as the number of cars $N$ on that route increases.

Initially, since $R_1$ is wider and hence faster, many drivers will choose that as their *best response*. This would increase its traffic load, i.e., congestion, which in turn increases the travel time in that route. This will discourage other drivers from switching to it and will encourage them to choose $R_2$ as their best response now. In this manner, drivers will self-organize and distribute themselves between the two alternatives, so that ultimately the travel time from home to work is the same in both routes, i.e., the *equilibrium* state. At equilibrium, there is *no incentive* for any driver to switch from his current route to the other, as the travel time is the same in both.

This is what we would expect to happen intuitively. Let us see whether the math bears this out. Given the problem description above, we have the following relations:

$$t_1 = \frac{N_1}{a_1} = \frac{x_1 N}{a_1}$$

$$t_2 = \frac{N_2}{a_2} = \frac{x_2 N}{a_2}$$

$$N_1 + N_2 = N$$

$$x_1 + x_2 = 1$$

where $t_1$ and $t_2$ are travel times on routes $R_1$ and $R_2$, respectively. Variables $N_1$ and $N_2$ are the number of cars on the two routes, $x_1$ and $x_2$ are fractional allocations of cars, and $N$ is the total number of cars on both routes.

The goal for every driver, of course, is to *minimize* her *travel time* $t$ from home to work. By minimizing the travel time, every driver is *maximizing* her *utility* or *welfare*. The travel time is the *cost* of the commute, and *negative cost* is the *utility* or the *benefit*. Therefore, we write the utility $h$ for the two routes as

$$h_1 = -\frac{x_1 N}{a_1}$$
$$h_2 = -\frac{x_2 N}{a_2}$$

As noted, in potential games, a player's payoff or utility is the gradient of potential $\phi(\mathbf{x})$, i.e.,

$$h_i(\mathbf{x}) \equiv \frac{\partial \phi(\mathbf{x})}{\partial x_i} \tag{3.7}$$

where $x_i = N_i/N$ and $\mathbf{x}$ is the population vector. Therefore, by integration [we replace the partial derivative with the total derivative because $h_i(\mathbf{x})$ can be reduced to $h_i(x_i)$], we have

$$\phi(\mathbf{x}) = \sum_{i=1}^{n} \int h_i(\mathbf{x}) \, dx_i \tag{3.8}$$

The potential $\phi(\mathbf{x})$ is therefore

$$\phi(\mathbf{x}) = -\frac{N}{2}\left(\frac{x_1^2}{a_1} + \frac{x_2^2}{a_2}\right)$$

We can show that $\phi(\mathbf{x})$ is strictly concave:

$$\frac{\partial^2 \phi(\mathbf{x})}{\partial x_i^2} = -\frac{N}{a_i} < 0 \tag{3.9}$$

Therefore, *a unique Nash equilibrium for this game exists*, where $\phi(\mathbf{x})$ is maximized, per the well-known theorem in potential games (Sandholm 2010, 60).

To determine the maximum potential, we use the method of Lagrange multipliers with $L$ as the Lagrangian and $\lambda$ as the Lagrange multiplier for the constraint $\sum_{i=1}^{n} x_i = 1$:

$$L = \phi + \lambda \left(1 - \sum_{i=1}^{n} x_i\right) \tag{3.10}$$

Solving $\partial L/\partial x_i = 0$, we obtain the following result at *equilibrium*:

$$h_1 = h_2 = \lambda$$
$$t_1 = t_2$$
$$x_1 = \frac{a_1}{a_2} x_2$$

We note that at equilibrium, the condition that $h_1 = h_2 = \lambda$, which is the same as $t_1 = t_2$, is exactly what was expected intuitively above—the travel times are the same in both routes. Furthermore, if both routes were equally wide, i.e., $a_1 = a_2$, then $x_1 = x_2$. So both routes will have the *same number of cars* as well, which is again what we would expect intuitively. So the math validates our intuition.

This is a useful result to explore further. You are told initially that there are two routes from home to work and that both are identical in all respects. You are also told there is a combined total of $N$ cars traveling in both routes. Then you are asked how the cars are distributed between the two routes, i.e., how many cars are traveling in route $R_1$ versus $R_2$. Since both routes are identical, there is no reason to prefer one over the other, and hence you would answer that the cars are distributed evenly. This would be the *fairest* distribution of cars between the two routes. There is no reason to assign more cars to one route and fewer cars to the other route. Clearly that wouldn't be fair.

Now, consider another scenario. You are told $R_1$ is twice as wide as $R_2$, or $a_1 = 2a_2$. You know this means $R_1$ is twice as fast as $R_2$, at least initially. That means it will attract more cars, which, of course, will slow it down. So, you'd reason that at equilibrium $R_1$ will have twice as many cars as $R_2$. This is the *fairest* distribution now. Keep these results in mind for now, particularly the relationship among a distribution, its fairness, and the problem constraints, for this connection is revisited in chapter 5.

This example is meant only as a simple illustration of best response dynamics in potential games, to give a nonexpert reader a quick introduction to the nature of the analysis. Interested readers are encouraged to see Easley and Kleinberg (2010) or Sandholm (2010) for a more thorough treatment of congestion games.

## 3.4 Is There an Equilibrium Income Distribution?

With this quick introduction to potential games, we now turn our attention back to the gedankenexperiment to answer this question: Is there an equilibrium income distribution?

Since our problem is a population game with features of a congestion game, we again use the potential game framework to address this question. Recall that an employee's utility is the gradient of potential $\phi(\mathbf{x})$, or

$$h_i(\mathbf{x}) \equiv \partial \phi(\mathbf{x})/\partial x_i$$

where $x_i = N_i/N$ and $\mathbf{x}$ is the population vector. Therefore, by integration [again, we replace the partial derivative with the total derivative because $h_i(\mathbf{x})$ can be reduced to $h_i(x_i)$, expressed in equations (3.1) through (3.4)]

$$\phi(\mathbf{x}) = \sum_{i=1}^{n} \int h_i(\mathbf{x}) dx_i$$

We observe that our game is a potential game with the potential function

$$\phi(\mathbf{x}) = \phi_u + \phi_v + \phi_w + \text{constant} \tag{3.11}$$

where

$$\phi_u = \alpha \sum_{i=1}^{n} x_i \ln S_i \tag{3.12}$$

$$\phi_v = -\beta \sum_{i=1}^{n} x_i (\ln S_i)^2 \tag{3.13}$$

$$\phi_w = \frac{\gamma}{N} \ln \frac{N!}{\prod_{i=1}^{n}(Nx_i)!} \tag{3.14}$$

where we have used Stirling's approximation in equation (3.14).

We can show that $\phi(\mathbf{x})$ is strictly concave:

$$\frac{\partial^2 \phi(\mathbf{x})}{\partial x_i^2} = -\frac{\gamma}{x_i} < 0 \qquad (3.15)$$

Therefore, as before, *a unique Nash equilibrium for this game exists*, where $\phi(\mathbf{x})$ is maximized.

Thus, the self-organizing free market dynamics, in which employees switch jobs and companies switch employees in search of better utilities or profits, ultimately reaches an equilibrium state, with an *equilibrium income distribution*. This happens when the potential $\phi(\mathbf{x})$ is maximized.

This immediately raises two questions: (1) What is the equilibrium income distribution? (2) What is the economic meaning of the potential function?

The first question is, in fact, the first question in the original set of four questions we raised in chapter 1. The second question here is new, but it turns out to be crucial also. We discuss and answer both questions in chapter 5.

## 3.5 Connection with Statistical Thermodynamics

Readers familiar with statistical mechanics will recognize the potential component $\phi_w$ as *entropy* (except for the missing Boltzmann constant $k_B$). This identification suggests that maximizing the payoff potential in game-theoretic equilibrium would correspond to *maximizing entropy* in *statistical mechanical equilibrium*, thereby revealing a deep and useful connection between these seemingly different conceptual frameworks. This connection suggests that one may view the statistical mechanics approach to molecular dynamics, also called *statistical thermodynamics*, from a potential game perspective. In this approach, one may view the molecules as restless agents in a game (let's call it the *thermodynamic game*), continually jumping from one energy state to another through intermolecular collisions. However, unlike employees who are continually driven to switch jobs in search of better utilities they desire, molecules are *not teleological*, i.e., *not goal-driven*, in their constant search. As prisoners of Newton's laws and conservation principles, constantly subjected to intermolecular collisions, their search and dynamical evolution are the result of thermal agitation.

We explore this connection in the next two chapters. In chapter 4, we first provide a simple, intuitive introduction to statistical thermodynamics and its key concepts, particularly entropy and statistical equilibrium. In chapter 5, we utilize this connection to derive formal and useful results regarding the income distribution in BhuVai.

CHAPTER FOUR

# Statistical Thermodynamics and Equilibrium Distribution

Everything should be made as simple as possible, but not simpler.
ALBERT EINSTEIN

---

Do not imagine that mathematics is harsh and crabbed, and repulsive to common sense. It is merely the etherealisation of common sense.
LORD KELVIN

## 4.1  Brief History

Statistical thermodynamics was born at the dawn of the industrial revolution out of the necessity to address challenges faced by engineers in the design and operation of efficient and reliable steam engines. They needed a predictive theory that could be verified experimentally, to help them make optimal decisions in the design and operation of complex energy and transportation systems driven by steam engines.

Over the course of about one hundred years, roughly from 1800 to 1900, this theoretical development was pioneered by great engineers and physicists such as Ludwig Boltzmann, Sadi Carnot, Rudolf Clausius, Josiah Willard Gibbs, Hermann Helmholtz, James Prescott Joule, Lord Kelvin, James Clerk Maxwell, Max Planck, William Rankine, Count Rumford, Julius Robert von Mayer, and James Watt (in alphabetical order). It is no exaggeration to state that the conceptual edifice they created is absolutely one of the most elegant and powerful intellectual achievements in human history. To date, it is the only successful theoretical framework we have, in any branch of science, that can predict system-level properties and dynamic behavior from entity-level ones, going from "parts to the whole," as it were.

In a sense, this is what we wish to accomplish in modeling the free-market dynamics of utility- or profit-maximizing teleological agents discussed in chapter 3, in developing a micro-to-macro economic framework. So the lessons from statistical thermodynamics are quite relevant to the problem we are addressing.

At first glance, though, it may seem far-fetched, even absurd, to look for guidance from a theory of inanimate systems comprising nonrational entities to develop a theory that needs to address the complex challenges in animate, teleological systems made of rational agents as seen in economics and sociology. However, a deeper analysis reveals surprising connections and useful results that provide fresh insights to this challenge.

The advancement of statistical thermodynamics required two important and exciting conceptual breakthroughs—the discovery of the *law of conservation of energy* and the discovery of the concept of *entropy*. The latter was particularly hard, involving an intellectual struggle that occurred in two stages. The first one lasted for about seven decades, from 1803 to 1877, as seen from the contributions of Clausius, Boltzmann, and Gibbs. The second conceptual struggle was between 1948, when Claude Shannon reintroduced entropy in the context of information theory, and 1957, when Edwin T. Jaynes reconciled both the thermodynamic and the information-theoretic interpretations.

The concept of entropy is somewhat mysterious to many, as Gibbs himself noted: "Any method involving the notion of entropy, the very existence of which depends on the second law of thermodynamics, will doubtless seem to many far-fetched, and may repel beginners as obscure and difficult of comprehension" (Gibbs 1906, 11). Even 150 years after its discovery, it is still one of the most misunderstood, and often maligned, concepts in science, requiring another makeover in the context of economics, as we shall show later in this chapter and the next.

For the sake of simplicity, we will illustrate only the key concepts and methodologies in the context of classical statistical thermodynamics and will ignore quantum effects. Our purpose here is to provide not a detailed introduction (which can be found in a number of excellent sources, such as Tolman 1938; Reif 1965; Nash 1970; Herbert 1985; Shankar 2014), but a quick summary of the essential conceptual elements of the framework.

In this book, we will use the term *statistical thermodynamics*, even though *statistical mechanics*, as the most general formulation of this approach, is the preferred term in physics. The reason we use *statistical thermodynamics* is to show the similarities and differences between thermodynamics and teleodynamics. In the former, the entities (e.g., molecules) are driven by thermal (i.e., heat) forces whereas in the latter the entities (e.g., people or firms) are driven by their goals (such as maximizing utility or profit). However, both systems reach statistical equilibrium by maximizing, as it turns out, surprisingly, the same quantity, as we prove in chapter 5.

## 4.2 "Molecular Society": Microstates, Macrostates, and Multiplicity

As described in chapter 2, any formal systems theory attempting to model a society of elementary entities needs to address the following essential aspects:

- *Microscopic view*: Define the elementary entities, their properties, their rules of interactions, and the microscopic states.
- *Macroscopic view*: Define the critical variables and parameters that can be observed and measured experimentally to determine the macroscopic state of the system.
- *Constraints*: Identify conservation laws or other such principles, if any, which constrain the system evolution and behavior.
- *Dynamics*: Identify the principles or rules that determine the evolution of the system over time.
- *Equilibrium*: Determine the system's statistical equilibrium, if it exists, and its stability.
- *Prediction*: Explain and predict macroscopic, system-level properties and behavior in terms of microscopic or entity-level properties.

For a *molecular society*, i.e., an organization of molecules in a given environment, the statistical mechanics framework accomplishes all this elegantly; it predicts with resounding success the macroscopic properties such as pressure, temperature, energy distribution, entropy, vapor-liquid equilibria, etc., given the microscopic properties of the molecules. Let us now summarize the key aspects of this extraordinarily beautiful and successful theory.

Let us start with an example of a thermodynamic system. There are three classes of systems in thermodynamics: *isolated* (both the total energy $E$ and the number of molecules $N$ are constant), *closed* (only the number of molecules $N$ is constant, and $E$ can vary due to exchanges with the surrounding environment), and *open* (both $N$ and $E$ vary due to exchanges with the environment). These arise in the context of an *ensemble*, introduced by J. Willard Gibbs, which is an idealization consisting of a large number of virtual copies of a system, considered all at once, each of which represents a possible state in which the real system might be located. The three important thermodynamic ensembles defined by Gibbs are the *microcanonical* ensemble (isolated system), *canonical* ensemble (closed system), and *grand canonical* ensemble (open system).

Consider a closed container with $N$ ideal monatomic gas molecules of the same kind, such as argon. The elementary entities of this "society" are argon molecules, all of which are endowed with the same properties that define an argon molecule. These are the mass, size, shape, position (three components for the three directions), velocity (also three components corresponding to the three directions), energy, etc. These may be considered their *inalienable* properties, as they define what a molecule is.

At this point, it is useful to differentiate between *intrinsic* properties and *extrinsic* properties. Properties such as mass, size, shape, and charge are the intrinsic properties that are absolutely the same for all the molecules; positions, momenta, and energies are extrinsic properties whose values can vary from molecule to molecule, as well as over time for the same molecule, due to intermolecular interactions. These molecular interactions as well as their movement or dynamics are governed by intermolecular collisions (as well as the molecular collisions with the walls of the container) subject to Newton's laws of motion, intermolecular forces of attraction and repulsion, and conservation principles. The molecules are thus prisoners of Newton's laws and conservation principles—they do not have their own "mind." Nor can they "decide" what to do next—they are not teleological. This is the microscopic or entity-level view of this society. A *microscopic state* of the system is defined as a particular configuration of the molecules that specifies the positions and velocities of all the molecules at any given time (and any internal states and quantum effects, which we ignore in our simple introduction).

In the macroscopic systems-level view, the entire molecular system, i.e., the molecular society, has macroscopic properties such as volume, pressure, temperature, energy, entropy, etc., which can be measured experimentally to determine the macroscopic state of the system. In general, there will be a very large number of microscopic states, *combinatorially* large, known as *multiplicity W*, that correspond to a given macroscopic state. The experimentally measured macroscopic properties are averages over such microscopic states. Multiplicity is given by Reif (1965)

$$W = \frac{N!}{N_1! \cdots N_n!} \qquad (4.1)$$

where $N_i$ is the number of indistinguishable molecules in category $i(i = 1, 2, \ldots, N)$, where $N$ is the total number of molecules. Essentially, this is the number of possible configurations the molecules could have in a given macrostate of a system. A macrostate that has a higher multiplicity is more likely to occur spontaneously than one with a lower one. This is the central idea behind the law of entropy, as we shall see.

The thermodynamic system evolves over time due to intermolecular interactions, and the macroscopic properties are determined by the molecular-level properties, molecular interactions, and environmental constraints. In due course of this evolution, if the system's macroscopic observables cease to change with time, then the system is said to have reached a *thermodynamic* or *statistical equilibrium*. For example, an ideal gas as an isolated system whose energy distribution has stabilized to a specific distribution, called the *Boltzmann distribution*, which is an exponential distribution, is said to be in thermodynamic equilibrium. This outcome allows a single temperature and pressure to be attributed to the whole system.

## 4.3 Phase Space, Entropy, and Statistical Equilibrium

In statistical thermodynamics, one analyzes the dynamics and evolution of the microstates of the system by using the concept of phase space. The *phase space* is an abstract multidimensional space of positions and momenta of all the molecules. This important representation facilitates probabilistic analysis that leads to experimentally testable macroscopic predictions. The phase space of $N$ monoatomic gas molecules in a closed container is the 6N-dimensional space (i.e., there are 6N coordinates) specifying the three positions and three momenta for each molecule (i.e., six coordinates per molecule). A point in this space specifies the positions and momenta of all the molecules, thus representing the microscopic state of the system at some time $t$.

In time, the system evolves and wanders all over the phase space due to intermolecular collisions and interactions governed by Newton's laws of motion and conservation laws. Typically, $N$ is an astronomically large number, about $10^{23}$ molecules, called the *Avogadro number*. This monstrous number, instead of being a hindrance, turns out to be a huge blessing in unleashing the power of probability theory and statistical analysis to make reliable macroscopic predictions.

The evolution of the molecular system is governed by two laws of thermodynamics (technically, there are four laws, but let's keep the discussion simple). They can be stated in a number of different ways, and here we use the simpler version. The first law, called the law of conservation of energy, states that energy can be neither created nor destroyed. The second law states that the entropy $S$ (not to be confused with salary) of an isolated system not in equilibrium will tend to increase over time, approaching a maximum value at equilibrium, where $S$ is defined by

$$S = -k_B \sum_i p_i \ln p_i \qquad (4.2)$$

where $p_i$ is the probability of finding the system in microstate $i$. The microscopic interpretation of entropy, as a measure of randomness or disorder, was provided by Boltzmann through his famous equation

$$S = k_B \ln W \qquad (4.3)$$

where $W$ is the multiplicity and $k_B$ is the Boltzmann constant. This is one of the most beautiful equations in physics and is inscribed in Boltzmann's tomb in Vienna.[1]

This conceptual interpretation of entropy as a measure of disorder was later generalized in an important way by Shannon (1948) as a *measure of uncertainty* in the context of information theory. This was later refined by Jaynes (1957a, b) in his formulation of the maximum entropy principle.

Right from the beginning, entropy has remained a somewhat confusing, even mysterious, concept. So what is entropy really? To answer this question, we need to look into the very origins of this concept 150 years ago, and why it was needed so badly in thermodynamics. Bear in mind that 150 years ago, the most successful, and hence the dominant, modeling framework was the Newtonian approach to classical mechanics and his laws of motion. In the Newtonian view, emphasis is on the individual entity, the "particle view," and its motion through space under the action of forces acting on it. It is not a "systems view" of the world, but an "entity view." This is, of course, most useful for certain kinds of problems, such as modeling planetary motion, but not for thermodynamics where there are about $10^{23}$ molecules dashing about.

The central puzzle in thermodynamics, in very simple terms, was the observation that heat "flows" *spontaneously* from a "hot" object to a "cold" object it is in contact with, but not the other way around; or, air at high pressure *spontaneously* rushes out of a balloon, when the balloon's mouth is untied, into the atmosphere, but not the other way around. Air doesn't spontaneously rush back into an empty balloon and build a high-pressure balloon.

To explain this further, consider the following simple experiment, shown in figure 4.1. Again, we wish to stress that this book is written mainly for economists and sociologists who may not have a background in statistical mechanics. Therefore, we avoid unnecessary technical details and complications, wherever possible, and try to elucidate the central concepts with simple examples.[2]

As in the balloon example, all the gas molecules are constrained in the left chamber by the partition in the middle, while the right chamber is completely empty. We know that when the partition is removed, the molecules will spontaneously rush into the empty chamber, and after some time, the distribution of gas molecules will look somewhat like that in figure 4.2, where the number of molecules in each chamber is more or less the same. This is what we refer to as *thermodynamic or statistical equilibrium*, when there is a *uniform* distribution of molecules; i.e., *an equal volume of space will have an equal number of molecules*, on average, anywhere in the two chambers. For illustrative purposes, we show only about 180 molecules in the figures. This distribution gets increasingly uniform, and the equality more precise, as the number of molecules increases. For real-life situations, where there are typically about $10^{23}$ molecules (e.g., in a typical party balloon filled with air), the equality is highly precise.

72  *Statistical Thermodynamics and Equilibrium Distribution*

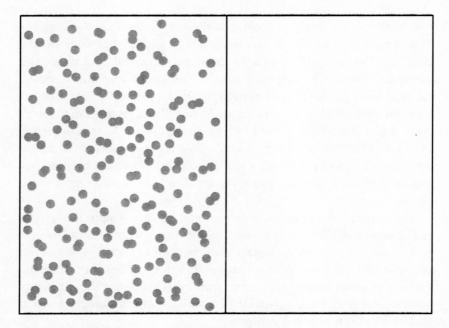

*Figure 4.1* All gas molecules are in the left chamber.

While we see the spontaneous transformation from figure 4.1 to 4.2 all the time, we never see the transformation 4.2 to 4.3 in practice, particularly when there are about $10^{23}$ molecules. That is, we don't find all the molecules spontaneously collapsing into one of the chambers. One will have to use some mechanism, such as a piston, and push all the molecules out of one chamber into the next.

It turns out such phenomena cannot be explained from a strictly Newtonian view of this system. What we mean by a Newtonian view is one in which one models this system as a collection of $N$ individual molecules, each one following Newton's laws of motion. One would then end up with $N$ second-order differential equations in position and time, one equation for each molecule.[3] These need to be solved to plot the individual trajectories of the molecules over time. For most real-life systems, $N$ is on the order of $10^{23}$, making it practically impossible to solve the equations. But even that is beside the point. Even if one could solve them, what one would get is the precise location of each molecule and its momentum, at any given time, and from these data

*Statistical Thermodynamics and Equilibrium Distribution* 73

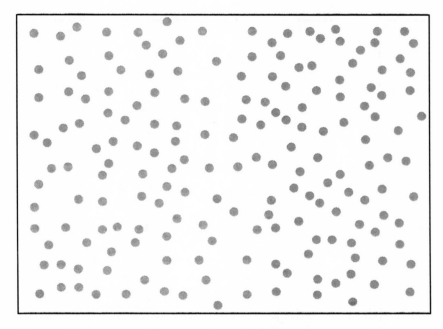

*Figure 4.2* Statistical equilibrium: uniform distribution of gas molecules.

one can plot the trajectory of each molecule. Even knowing all this detailed information about each molecule doesn't help us explain why the transformation 4.2 to 4.3 doesn't happen spontaneously. In the Newtonian picture, this transformation is fully allowed by the dynamic equations. It is the same puzzle regarding the spontaneous flow of heat from a hot object to a cold one, but not the other way around.

Thus, this was a huge intellectual challenge requiring a major conceptual breakthrough. After a monumental conceptual struggle lasting about six decades, at least from 1803 to 1867, Rudolf Clausius (1867) finally identified this missing concept, defined it mathematically, and gave it the name *entropy*. He had been refining his intuition about this concept from 1854, when he first wrote about it, calling it *equivalence-value*. In his own words, he wrote

> I propose to name the quantity $S$ the entropy of the system, after the Greek word [τροπη trope], the transformation. I have intentionally formed the word *entropy* so as to be as similar as possible to the word

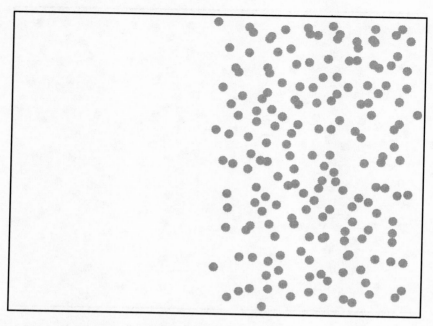

*Figure 4.3* An extremely unlikely configuration.

*energy*: for the two magnitudes to be denoted by these words are so nearly allied in physical meanings, that a certain similarity in designation appears to be desirable. (Clausius, 1867, 357; see also Simpson et al. 1989).

Although Clausius did not mention why he chose the symbol $S$ to represent entropy, it is believed that he chose it in honor of Sadi Carnot, the great French engineer who was the first leading researcher in thermodynamics and often considered the "father of thermodynamics."

When Clausius introduced his definition, he formulated it in the macroscopic thermodynamic form $dS = dQ/T$, where $Q$ is the amount of heat energy and $T$ is the temperature. It was left to Ludwig Boltzmann, in 1877, to provide a molecular or microscopic interpretation of entropy as a measure of the microstates accessible to a dynamic system in a given macrostate, i.e., multiplicity $W$ [equation (4.3)].

This is what happens in our experiment in figures 4.1 to 4.3. By removing the partition in figure 4.1, we have increased *multiplicity*, by making additional molecular configurations (i.e., microstates) accessible to the system in the newly opened up space. In other words, the entire right chamber is now accessible to all the molecules to diffuse into. Thus, the multiplicity explodes in value, making it the *overwhelmingly likely* outcome in the long run, i.e., thermodynamic or statistical equilibrium. In our example, for $N = 180$, at equilibrium there are about 90 molecules, on average, in the left half and another 90 in the right half. So the multiplicity $W = 180!/(90!90!) \sim 10^{52}$, a very large number. That is, there are about $10^{52}$ microstates for the equilibrium macrostate—there are about $10^{52}$ ways of reaching the equilibrium configuration. Any one of them will do.

Contrast this situation with the case where all the molecules retreat into the right chamber (or for that matter, back into the left chamber), as in figure 4.3. In this case, multiplicity collapses to *exactly* 1, because there is exactly one configuration that gets you there. This makes it an *extremely unlikely* macrostate to occur spontaneously, whereas figure 4.2 is *extremely likely* to occur spontaneously. In real-life situations, where $N$ is on the order of $10^{23}$, not 180, the multiplicity for the equilibrium macrostate is on the order of $10^{10^{23}}$, an unimaginably large number of gargantuan magnitude, guaranteeing that the equilibrium macrostate is not one of probability but one of virtual certainty! Thus, maximizing entropy is equivalent to determining the *most probable outcome* eventually, i.e., the equilibrium macrostate.

Thus, the puzzle was finally solved by invoking a *system-level concept*, entropy, which was missing in the Newtonian picture, as it was focused on individual entities, not the system. It was missing the forest for the trees, as it were. *Entropy is a systems property*; a molecule doesn't have entropy, but a collection of molecules does. Thus, statistical thermodynamics was born as the first systems theory that successfully bridged the parts-to-whole gulf, going from micro to macro.

In this simple example, we see the connection among statistical equilibrium, entropy, and the equality of all accessible space. Let us explore this connection further. Imagine now that the entire physical space in our apparatus is divided into extremely small cells of volume $\delta v$ (say, 1 nanoliter, i.e., one-billionth of a liter). Let's call them *unit cells*. If our apparatus has a volume of 10 liters (the equivalent of five 2-liter soda bottles), then there are 10 billion ($10^{10}$) unit cells. We assume that the total number of molecules vastly outnumbers the number of unit cells. (This is opposed to what is shown in figure 4.1, where there are only about 180 molecules. In reality, this would be around $10^{23}$, as noted.)

Let $\delta N$ be the number of molecules found in a unit cell at any given time. It will, on average, be around $\sim 2 \times 10^{13}$ ($= 10^{23}/0.5 \times 10^{10}$), in the left chamber in figure 4.1. Call this the *occupancy number*. It won't, of course, be the same molecules all the time, as the system is highly dynamic with all the molecules rapidly flying about, at speeds of $\sim 1100$ miles per hour, colliding into one another at the rate of $\sim 10^9$ collisions per second. So the occupancy number will fluctuate somewhat over time, given the stochastic nature of the underlying collision process, but it will nevertheless remain remarkably close to the average value for all practical purposes. In figure 4.1, only the unit cells in the left chamber are *accessible* to the molecules, and the ones in the right are not accessible because of the partition. So, on average, all the unit cells in the left chamber are likely to be occupied by $\delta N$ molecules (or $\sim 2 \times 10^{13}$), at any given time at equilibrium.

Once the partition is removed, the molecules will rush into the right chamber, as those unit cells have now become accessible to them. Soon a new statistical equilibrium will set in, when all the unit cells in the right chamber will have, on average, the same (but new) occupancy number ($10^{23}/10^{10} = 10^{13}$) as the ones in the left, except for the statistical fluctuations, which are negligible. The crucial point is that at equilibrium *all cells*, in both chambers, will have the *same* occupancy number. This is, in fact, the defining criterion for statistical equilibrium—all cells are equal with respect to occupancy. In other words, a given molecule is equally likely to be found in any one of the accessible physical space cells at any given time. This observation can be easily generalized from accessible physical space cells to accessible phase space cells.

This defining criterion of statistical equilibrium is, in fact, the *fundamental postulate* of statistical mechanics. Known as the *equal a priori probability postulate*, it states: "In equilibrium, every allowed microstate of an isolated system has the same probability" (Shankar 2014, 437).

Now that we have seen the central concepts of statistical thermodynamics, even though only briefly, we are ready to proceed to the next step of generalizing this framework for teleological systems. Toward the end of chapter 3, we observed that the game-theoretic potential $\phi_w$ is the same as entropy in statistical thermodynamics. We also proved that the game-theoretic Nash equilibrium (i.e., the equilibrium distribution of salaries in our gedankenexperiment) is reached when $\phi$ is maximized. In this chapter, we showed how statistical equilibrium is reached when entropy is maximized.

So is there a connection between these two equilibria? In what way do the concepts and techniques of statistical thermodynamics help us address the question of what is the fairest inequality we ought to see in a free-market environment? These are the questions we answer in chapter 5.

CHAPTER FIVE

# Fairness in Income Distribution

> Nature uses only the longest threads to weave her patterns, so each small piece of her fabric reveals the organization of the entire tapestry.
> RICHARD FEYNMAN

> Otherwise seeing he sees not, hearing he follows not. But to him (to the qualified), she (Vak in the form of speech) reveals her form as a loving well-dressed wife disrobes herself to her husband.
> RIG VEDA 10.71-4

Statistical mechanics is the conceptual framework for computing the macroscopic properties of a thermodynamic system from its microscopic ones. It also provides guidance in identifying the essential components of a formal framework for modeling systems of many interacting entities. Thus, it gives us a starting point for developing such a framework for modeling teleological systems—systems of goal-directed entities or agents. In this chapter, we develop this new framework, named *statistical teleodynamics*. We start with a brief introduction to complex teleological systems. Then we discuss statistical teleodynamics, its relationship to the game-theoretic formulation we developed in chapter 3, and the concept of fairness in income distribution.

## 5.1 Complex Teleological Systems

Teleological agents are *rational* entities with a *purpose* or *goal*, unlike molecules of a gas, which are not driven by a purpose. The molecules are mere prisoners of Newton's laws and conservation principles. They do not have a "mind" or "purpose" of their own to decide on a future course of actions. Teleological agents, however, wish to maximize their individual utilities, and take a course of actions that would help them accomplish this objective. The employees in a corporation, for instance, are thus teleological. The corporation itself is teleological as well, with its own purpose of surviving and growing in a competitive business environment by maximizing profit.

Teleological systems may be *human-engineered*, such as the ones in engineering (e.g., Internet, transportation networks, national power grids), in economics (e.g., corporations, supply chain networks), and in sociology (e.g., government organizations); or *naturally evolved* such as in biology (e.g., cellular and metabolic networks, food webs, ecosystems) and in economics and sociology (e.g., human societies and the free market, although these are also human-engineered to some extent).

One may view the primary purpose of teleological systems as survival and growth in their given environment. For example, in engineering, a power grid or a chemical plant has built-in algorithms and mechanisms to make the system meet certain performance criteria and to shut itself down in the case of abnormal situations. In biology, an organism needs to execute the functions of reproduction and growth, at some desired performance levels dictated by the environment, in order to survive, grow, and propagate its species. In economics, a corporation needs to execute a variety of business and/or manufacturing functions efficiently, safely, and quickly in order to survive and grow in a competitive market for its customers. Thus, the overall purpose is survival-related, always in the context of an operating environment.

The environment determines what functions or features, and their target levels of performance, are expected of a teleological system in order to survive and grow in that environment. For instance, in his famous study of finches in the Galapagos Islands, Charles Darwin (1859) noted that the size and shape of the birds' beaks were strongly influenced by environmental factors. In economics, 10% may be an acceptable return on investment (ROI) in one industry whereas it might be 20% in some other. Yet another industrial segment might disregard ROI altogether and evaluate companies based on their market share growth. Thus, the environment imposes both qualitative (e.g., ROI, market share) and quantitative constraints (e.g., 10%, 20%) on the design of teleological systems. These determine the overall level of fitness a given system must possess to survive in its environment. The survival-growth objective differentiates teleological systems from arbitrary thermodynamic systems.

In addition, there is another design criterion imposed by the ever-changing dynamic environment. One would like the system to be designed in such a manner that it could survive and meet its performance targets under a wide variety of potential operating conditions in the environment. For example, in the design of computer and communication networks, engineers try to first anticipate a wide variety of communication patterns, demand scenarios, failure of nodes and edges, addition of new nodes and edges, and so on, and then devise a network that can survive under such varied operating conditions and meet its performance target criteria. A reliable design would not presume that only certain conditions will prevail in the future and thus limit the design to perform well only under those conditions. That is, one would not wish to *bias* the design toward a specific operating environment, particularly if the nature of future operating environments were uncertain.

Hence, the system design should reflect this inherent uncertainty about future operating environments and minimize the bias or any unwarranted assumptions about them. In other words, a reliable design should minimize potential risk by accommodating as much uncertainty about the future operating environments as allowed by the constraints, i.e., reliability under maximum uncertainty. We refer to this criterion as *optimally robust design*. This concept is obviously related to the stability of the system under various disturbances, and it is therefore an important criterion for the stability of human societies and the free market as well. We examine this implication in greater detail in section 5.9.

## 5.2 Statistical Teleodynamics: Intrinsic and Extrinsic Properties

Exploring the agent-level perspective, we need to identify the essential properties, intrinsic and extrinsic, and their associated values (i.e., magnitudes), which define an employee agent. The *intrinsic* properties of an employee are innate attributes that are exactly the same for all employees. These are the fundamental rights, such as the right to life and liberty, to a discrimination-free, healthy work environment, etc. These ought to be guaranteed to be absolutely the same for all the employees, and all employees are to be treated equally in this regard. This requirement in our theory corresponds to the first principle, the liberty principle, in the Rawlsian framework, where every individual in a civil society has equal basic liberties. Rawls arrives at this fundamental principle of equality through the application of his "veil of ignorance" concept to a group of rational agents in the original position. In our theory, we arrive at this by using the maximum entropy principle, which we discuss in chapter 8. So, for the moment, let us assume this following Rawl's justification.

As for the *extrinsic* properties, these properties are typically acquired through education, training, experience, etc., even though some can be a result of genetic inheritance, such as one's appearance, mental or athletic abilities, or personality traits. These can determine the performance level of the employees in meeting the teleological objectives of the system. These can vary from employee to employee and, unlike intrinsic properties, are not guaranteed to be equal. Some of these may be relevant to job performance (such as education and experience), and others may not be (such as one's skin color, gender, or race). Salary, e.g., is an extrinsic property. Every employee is guaranteed a salary, but the amount can be different for different employees based on their contributions to the company's overall success. Again, in a comparison with molecules of a pure monoatomic gas, the mass, size, etc., are intrinsic properties and thus are absolutely the same for all the molecules, while positions, momenta, and energies are extrinsic properties, which are different for different molecules.

At the system level, the extrinsic properties would be macroscopic properties such as revenue, expenses, salary budget, profit, number of employees, and market share. Again, by invoking the thermodynamic analogy, these would be akin to pressure, temperature, volume, number of molecules, total energy, entropy, etc. However, there is *no one-to-one correspondence* implied between these two lists of macroscopic properties.

## 5.3   Phase Space in Statistical Teleodynamics

Borrowing from the phase space representation in statistical thermodynamics described in sections 4.2 and 4.3, we can define the phase space of iAvatars Inc. as an $N$-dimensional space with each dimension representing the salary $S_j$ of an employee $j$. In reality, this space could be more complicated with additional dimensions that capture other features such as titles, awards, etc., but let us restrict ourselves to salary, since it is the dominant component of a person's utility, as discussed in chapter 3. A point in this ideal phase space represents the microstate of the company at some time $t$ by specifying the salaries of all the employees. iAvatars Inc. is an isolated system, in the thermodynamic sense, with respect to $N$ and $M$ (i.e., these are constant) but an open system with respect to information exchange regarding the salaries and job openings elsewhere in the free-market economic environment.

The role of information exchange is an important distinction between thermodynamic systems and economic teleodynamic systems. In thermodynamics, as noted, we have three classes of systems: *isolated* (both total energy $E$ and the number of molecules $N$ are constant), *closed* (only the number of molecules $N$ is constant, and $E$ can vary due to exchanges with the surrounding environment), and *open* (both $N$ and $E$ vary due to exchanges with the environment). Thus, there are only two kinds of exchanges: *energy* and *matter*. However, in this free-market environment, we have one more exchange mechanism due to information exchange. We can have a situation in which a system is isolated with respect to money and employees (i.e., these are constant as they are in our case) but open with respect to information exchange. Note, however, that information exchange is not subject to conservation laws like those of energy and matter.

Now consider an isolated thermodynamic system of $N$ gas molecules and a total energy $E$, with an average energy $E_{ave} = E/N$. In such an isolated system, both $N$ and $E$ are conserved. Imagine starting this system at $t = 0$ in the initial state, where all molecules are assigned to have exactly the same energy $E_{ave}$, as we did for the salaries in our thought experiment in section 3.1. In time, the system will evolve from this initial state of a Kronecker delta distribution in energy, wander all over the phase space driven by intermolecular collisions that cause energy exchanges among the molecules, to one where the energy distribution is more spread out, similar to what happened to the salary distribution. We know from statistical mechanics that, as we discussed in chapter 4, this spreading is related to increasing entropy. Eventually this spreading will settle down in an equilibrium distribution, the well-known *exponential Boltzmann distribution*, when entropy is maximized.

In a similar manner, as the salaries of the employees increase and decrease in our thought experiment (see figure 3.1), the company iAvatars Inc. adapts, evolves, and wanders through its phase space. Of course, the mechanistic details of the evolution of the two systems are very different. For molecules, the system evolves through molecular collisions governed by Newton's laws of motion and conservation laws. For employees, it is through the exchange of information among the employees, company, and free-market environment, driven by the desires of the employees to maximize utility and those of the company to maximize its profit. Information exchanges are the "collisions" in this regard.

## 5.4 What Is the Equilibrium Income Distribution in BhuVai?

As you may recall, in section 3.3, we proved that our thought experiment results in an equilibrium distribution of salaries and that it is a Nash equilibrium. We also observed, in section 3.4, that this implies that the statistical thermodynamic equilibrium reached by colliding gas molecules can also be modeled as a potential game resulting in a Nash equilibrium. We follow up on this intriguing connection first, before answering the question regarding the equilibrium income distribution. We, therefore, first answer the question "What is the equilibrium energy distribution?" for the thermodynamic game, namely, the dynamic evolution of molecular microstates and the resultant thermodynamic macrostate.

Approaching the thermodynamic game from the potential game perspective, we have the following "utility" for molecules in state $i$:

$$h_i(E_i, N_i) = -\beta E_i - \ln N_i \quad (5.1)$$

where $E_i$ [not to be confused with effort[1] in equation (3.5)] is the energy of a molecule in state $i$, $\beta = 1/(k_B T)$, $k_B = 1.38 \times 10^{-23}$ JK$^{-1}$ is the Boltzmann constant; and $T$ is temperature.

By integrating the utility, we can obtain the potential of the thermodynamic game:

$$\phi(\mathbf{x}) = -\frac{\beta}{N} E + \frac{1}{N} \ln \frac{N!}{\prod_{i=1}^{n}(Nx_i)!} \quad (5.2)$$

where $E = N\sum_{i=1}^{n} x_i E_i$ is the total energy that is conserved.

We use the method of Lagrange multipliers with $L$ as the Lagrangian and $\lambda$ as the Lagrange multiplier for the constraint $\sum_{i=1}^{n} x_i = 1$:

$$L = \phi + \lambda \left(1 - \sum_{i=1}^{n} x_i\right) \quad (5.3)$$

Solving $\partial L / \partial x_i = 0$ and substituting the results back into $\sum_{i=1}^{n} x_i = 1$, we obtain the well-known *Boltzmann exponential distribution* at equilibrium:

$$x_i = \frac{\exp(-\beta E_i)}{\sum_{j=1}^{n} \exp(-\beta E_j)} \quad (5.4)$$

What we just did is the standard procedure followed in maximum entropy methods in statistical thermodynamics and information theory to identify the distribution that maximizes entropy under the given constraints (Jaynes 1957a, b; Kapur 1989).

Once again, readers familiar with statistical thermodynamics will recognize that from equation (5.2)

$$\phi = -\frac{1}{N k_B T}(E - TS) = -\frac{\beta}{N} A \quad (5.5)$$

where $A = E - TS$ is the Helmholtz free energy, $S$ is entropy (again, not to be confused with salary), and $T$ is temperature. Indeed, in statistical thermodynamics $A$ is called a *thermodynamic potential*. In this regard, we could also see the correspondence between utility $h_i$ and the *chemical potential*, the partial molar free energy.

These valuable results, reproducing well-known equations in statistical thermodynamics, validate the soundness of our game-theoretic framework and its direct correspondence with statistical thermodynamics. Maximizing the game-theoretic potential is the same as maximizing entropy, for they are one and the same, except for the Boltzmann-constant multiplier.

Following this logic, for the *teleodynamic game* (i.e., the income distribution game), we carry out the same procedure to maximize $\phi(\mathbf{x})$ in equations (3.11) through (3.14) to obtain the following *lognormal distribution* at equilibrium:

$$x_i = \frac{1}{S_i D} \exp\left\{-\frac{[\ln S_i - (\alpha + \gamma)/2\beta]^2}{\gamma/\beta}\right\} \quad (5.6)$$

where $D = N\exp\left[\lambda/\gamma - (\alpha+\gamma)^2/4\beta\gamma\right]$ and $\lambda$ is the Lagrange multiplier.

Note that even though our analysis is analogous to that in statistical thermodynamics, the resulting distribution is different from that of the ideal gas. In the case of ideal gas molecules, the corresponding energy distribution is exponential, whereas the income distribution is lognormal. The difference is due to the differences in the objective functions and the constraints between the two systems.

Furthermore, it is important to recognize that while there is a close relationship between statistical thermodynamics and statistical teleodynamics, it is not appropriate to look for one-to-one correspondences for all kinds of properties and relations. For example, it is meaningless to look for the exact equivalents of pressure, volume, and temperature, or the *Maxwell relations*, in statistical teleodynamics. One might use certain terms loosely, such as competitive "pressure" and a "hot" market, but these are not to be taken literally. In this regard, we agree with Samuelson's objection mentioned earlier. The connection between the two disciplines occurs at a deep, fundamental level (as we discuss in section 5.6) and not through such superficial correspondences.

We recall, from chapter 3, that the spreading of the income distribution is driven by notions of fairness (or unfairness). But from the statistical thermodynamic, or the statistical teleodynamic, arguments we see that the spreading is driven by increasing entropy.

How do we reconcile these two interpretations? What is the connection between entropy and fairness? This is what we explore next.

## 5.5   Is the Equilibrium Distribution Fair?

Let us revisit our thought experiment in chapter 3. All employees in iAvatars Inc. contribute to its overall success in their own ways. How do we value each one's contribution and reward them all fairly? That is, what is the fairest distribution of salaries?

Let us now consider two extreme compensation scenarios. In scenario 1, as we discussed earlier, all employees are treated equally and are paid the same salary. Obviously, disenchanted employees will leave, and there will be no company left, despite its noble intentions. In scenario 2, only the CEO is considered to be most valuable, and everyone else is of minimal value. So the CEO gets most of the profits as her pay, and the rest of the employees are paid negligible amounts. Obviously, employees will flee the company in droves, and soon there will be no company left. Clearly, the reality lies somewhere between these two extremes of valuation schemes. But where?

According to Rawls (2001), from his *Justice as Fairness* theory, the fairest wage distribution maximizes the salary of the lowest-paid employee, which is the maximin strategy, enforced by some central authority, perhaps the board of directors in this case, or even the state through a minimum-wage policy. But Rawls's theory does not answer the question of what this distribution is, or whether we might have an equilibrium distribution, even under ideal conditions.

If we follow Nozick, his prescription is not to let some central authority decide what is fair, but to let the free market work its magic. According to him, the fairest income distribution is arrived at *naturally* under the guidance of the "invisible hand," resulting from the self-organizing dynamics of the free market. In a competitive free market environment, companies and employees as rational agents arrive at this distribution iteratively, by a trial-and-error evolutionary feedback process, through the free exchange of information and people among companies and the free-market environment until equilibrium is reached. That is, the equilibrium distribution is "discovered" through such evolutionary adaptation. Like Rawls, Nozick also does not explain what this distribution is, even under ideal conditions.

While Rawls and Nozick cannot inform us what this distribution is quantitatively, we do know what it is for the ideal free market in our hybrid utopia. As we just proved, it is the lognormal distribution.

But is this income distribution fair?

### 5.5.1  Fairness: Three Fundamental Principles

Before we answer this question, let us examine what we mean by *fair*. Our sense of fairness is founded on three fundamental principles:

1. *Equality principle*: This may be stated in a number of ways, but they all essentially express this concept: in a fair decision process one must not treat two equal entities unequally, e.g., there must be equal pay for equal work.
2. *Proportionality principle*: This principle also can be stated in different ways, depending on the context, but again, all essentially express the notions "you get more for doing more" and "you get

less for doing less," in a fair decision process. For example, "punishment should fit the crime" is the foundational principle of a fair judicial process in civil societies. One gets a larger punishment for committing a bigger crime, and a smaller one for smaller bad acts. Another example is "more pay for more work." This principle is essentially a variation of the equality principle, properly scaled. For instance, if John makes $100 for one hour of work, then Lilly should be paid $200 for doing two hours of the same work. That is, they are compensated equally on a properly scaled basis. However, in general, this principle does not necessarily imply a strict linearity—it merely implies that "you get more for doing more" and "you get less for doing less."

3. *Arbitrariness avoidance principle*: This principle states that in a fair decision process, decisions are not made arbitrarily, but are based on facts in evidence. For example, if there were two equally qualified candidates for a job, and if one rejected one candidate arbitrarily over the other, we would consider that an unfair treatment. Thus, rejecting or accepting a particular option, among many other options, for no reason would be considered unfair. Our sense of fairness requires that all options be considered, evaluated based on the merits and demerits of each option if such information is available, and then a decision be made. This requirement also is a variant of the equality principle—arbitrarily choosing one candidate over another equally qualified candidate violates the equality principle. This requirement is also a restatement of the principle of insufficient reason, as we discuss below. Even though this awkwardly named principle is already implied in the equality principle, it is nevertheless useful to highlight the importance of avoiding arbitrariness in decision-making.

Thus, if we examine fairness closely, we find that the equality principle is at the very core of our notion of fairness. Therefore, if there are $N$ possible candidates among whom a resource is to be distributed, and if one has no information for preferring one candidate over another, then the fairest distribution of the resource is equal allocation among all of them. This quantitative mathematical relationship expressing equality is at the core of the concept of fairness.

Jaynes (1979) points out that the notion of equal allocation was perhaps first expressed mathematically by Gerolamo Cardano, around

1564, in his book *Liber de Ludo Aleae* (*The Book on Games of Chance*). Jacob Bernoulli (1713), in his seminal work *Ars Conjectandi*, and Pierre Simon Laplace (de Laplace 1814, 12), in *Essai philosophique sur les probabilités*, expressed this notion while laying the foundations of probability theory, as the *principle of insufficient reason*. Laplace wrote:

> The theory of chance consists in reducing all the events of the same kind to a certain number of cases equally possible, that is to say, to such as we may be equally undecided about in regard to their existence, and in determining the number of cases favorable to the event whose probability is sought. The ratio of this number to that of all the cases possible is the measure of this probability, which is thus simply a fraction whose numerator is the number of favorable cases and whose denominator is the number of all the cases possible.

This is a fundamental principle but with an awkward name which, as Jaynes observes (Jaynes 1979, 2), "has had, ever since, a psychologically repellent quality that prevents many from seeing the positive merit of the idea itself." This required a makeover, which was done by the great economist John Maynard Keynes (1921) who renamed it the *principle of indifference*.

### 5.5.2 Entropy, Multiplicity, and Fairness

It is worth exploring the close connection among these three concepts—entropy, multiplicity, and fairness—in the context of teleological systems, to clarify certain subtle aspects. While the following discussion may seem unnecessary for experts in probability theory and statistical mechanics, we believe it will be helpful to others. We are deliberately considering a simple example so that readers can more easily see and work out the details of this connection.

Consider an artisan sweet shop, Ananda Malai 108, LLC, that makes blissfully delightful confections, known for its large assortment of artisan sweets and pastries with increasing degrees of sophistication in the ingredients, appearance, and taste. The shop employs eight talented workers (i.e., $N = 8$), all of whom are equally skilled in a variety of confectionery tasks. The shop has four categories of tasks (i.e., $n = 4$) of increasing order of difficulty and sophistication. An artisan confection would require the execution of some or all categories to varying extents depending on the sophistication of the final product. The more sophisticated the sweet, the more valuable it is (and so more pricey); hence greater effort is required to create it. All categories contribute value— $V_1 = \$100$, $V_2 = \$200$, $V_3 = \$300$, $V_4 = \$400$ per employee per day— toward the final value $V$ of the finished product. The job categories are essentially value creation categories. The shop's revenue comes from a combination of custom orders and walk-in sales; both are somewhat hard to forecast. They carry minimal inventory to keep things fresh.

Every day, when they arrive for work, the eight employees are assigned to different categories depending on the workload that day. All are equally skilled in all four categories, and so they prefer to be rotated through different categories on different days to enjoy some variety in their work. Also the categories are progressively difficult and stressful, demanding more out of the employees even though they get paid more. So the employees prefer lighter assignments a couple of times a week. All employees get to keep 50% of the value they create; e.g., an employee would make $50 per day in category 1, $100 per day in category 2, and so on. The remaining 50% covers the cost of ingredients, overhead, and profit for the owner.

Once assigned upon arrival, the employees start their work, and it is expensive to change their assignments in the middle of their tasks as it leads to wastage of material, equipment, and time. On most days, the workload is generally known at the beginning of the day based on the orders that came in the day before. However, on some days the workload is not known at the start of the day, and it takes a couple of hours for the orders to be known.

SCENARIO 1: UNKNOWN DEMAND Let us now consider different demand scenarios. In scenario 1, the workload is not known for the day. Under this condition, how should the eight employees be assigned among the four categories?

There are a large number of possibilities: [8 0 0 0], [0 8 0 0], [0 0 8 0], [0 0 0 8], [7 1 0 0], [5 2 1 0], [2 2 2 2], [4 3 0 1], [3 3 1 1], [1 2 3 2], [2 2 0 4], etc., where the format of this representation is the number of employees in each category in the following order: [category $-1$ category $-2$ category $-3$ category $-4$].

Now what is the fairest assignment of the employees among the categories? In other words, based on the given information, how do we treat all *options* (i.e., categories) fairly?

Consider the assignment [8 0 0 0]. It presumes that on that day only category $-1$ tasks will be needed and none of the other categories will be required. However, there is *no basis* for this decision. It *arbitrarily* rejects the other categories, violating the arbitrariness avoidance principle. We call this *structural arbitrariness*, where the structure of the assignment or the model is decided arbitrarily. Now consider [3 3 1 1]. This accommodates all categories, and in that sense it is better than [8 0 0 0], since it does not rule out any category arbitrarily. However, it assigns arbitrarily the *number* of employees to different categories. Why should category $-1$ and category $-2$ have three each and one each for the other two? What is the basis for this assignment? Thus, this also violates the arbitrariness avoidance principle. We call this version *parametric arbitrariness*, where the structure is fine but the parameters are decided arbitrarily.

The fairness principles require that the assignment be as *broad* as possible in including categories (i.e., not reject any feasible category arbitrarily) and as *uniform* as possible in their magnitudes (i.e., not reduce or increase the magnitude of a feasible category arbitrarily), while accounting for all the known information without assuming anything other than what is given. Hence, the fairest distribution is the broadest and the most uniform distribution that is allowed by the constraints. Since there are no constraints other than $N = 8$ and $n = 4$, the fairest distribution in this case is the uniform distribution [2 2 2 2]. This is the only distribution that is justified by the given information. This distribution does not assume anything beyond what is given.

Now the fairest distribution is also the best assignment from a business perspective. Again, consider [8 0 0 0]. It presumes that only the simplest confection orders will come in that day. It's a brilliant choice if that's what happened that day. But what if orders for more sophisticated sweets arrive and none arrive for the simpler confections? The shop will not be able to meet the demand as it didn't assign employees to the other categories. Since there are thousands of potential variations in sweets orders, it is extremely unlikely that only the simplest sweets will be ordered that day. Therefore, it is a poor business decision to bet on that. Similarly, consider [0 0 0 8]. This has presumed the opposite—that only the high-value sweets will be ordered. Again, it is a poor decision. The same logic applies to all the other choices to varying degrees except [2 2 2 2]. Some are really bad decisions, such as the ones we considered, and others are less so (e.g., [3 3 1 1] or [1 2 3 2]) as they are somewhat better prepared.

The assignment [2 2 2 2] is the optimal decision under the circumstances, as it provides maximum flexibility and maximum preparedness. It has not assumed any particular order flow—in that sense, it is the least-biased distribution. Therefore, maximizing fairness is maximizing the robustness of the operation by minimizing the risk of wrongly guessing the order flow (Venkatasubramanian 2007). (We return to this point at greater length in sections 5.9 and 8.2.) Thus, the fairest assignment is the optimal assignment from a business perspective. The owner might have preferred hiring four (or more) additional employees and having the distribution as [3 3 3 3], which would have enabled him to be even more prepared to face an uncertain demand, but his budget constraint does not allow him this capability. Thus, [2 2 2 2] is the best he can do given that his budget constraint limits $N$ to 8.

SCENARIO 2: KNOWN AGGREGATE DEMAND Let us now consider a different demand scenario. A customer called and paid for $2400 (i.e., $V = 2400$) worth of sweets, but she hadn't quite made up her mind about what combination of sweets she wanted the shop to make. She had agreed to let them know in a couple of hours after they got started with the preliminary preparations.

So what is the best option from a business perspective? Consider the assignment [0 0 8 0]. This will produce eight $300 confections worth $2400. This satisfies the constraint, but what if she wants an assortment of sweets, some simple, others more sophisticated, for a total of $2400? If the shop produces only $300 sweets, this will not satisfy her needs. So this assignment also makes an unwarranted assumption, as in the previous scenario, about the nature of the demand. This assumption is not supported by the information provided.

Again, the owner needs an assignment that will give him maximum flexibility, maximum preparedness, to be ready to deliver on an uncertain demand. Thus he needs to staff all the task categories as broadly and as uniformly as possible within the space of alternatives allowed by the constraint.

TABLE 5.1
Feasible Assignments and Their Multiplicities

| Category | 1 | 2 | 3 | 4 | 5 | 6 | 7 | 8 | 9 |
|---|---|---|---|---|---|---|---|---|---|
| 1 | 0 | 0 | 0 | 2 | 1 | 0 | 2 | 1 | 1 |
| 2 | 0 | 1 | 4 | 1 | 0 | 2 | 0 | 2 | 1 |
| 3 | 8 | 6 | 0 | 0 | 5 | 4 | 2 | 1 | 3 |
| 4 | 0 | 1 | 4 | 5 | 2 | 2 | 4 | 4 | 3 |
| Multiplicity $W$ | 1 | 56 | 70 | 168 | 168 | 420 | 420 | 840 | 1120 |

The alternatives are listed in table 5.1. Recall from equation (4.1) that the multiplicity of an assignment is given by

$$W = \frac{N!}{N_1! \cdots N_n!}$$

where $N_i$ is the number of employees in category $i$ ($i = 1, 2, \ldots, n$), and $N$ is the total number of employees. In our example, $N = 8$ and $n = 4$. Table 5.1 shows all the assignments that satisfy the \$2400 constraint and their multiplicities. The multiplicity is the number of different ways we can distribute the employees for a given assignment. For example, there is only one way of distributing employees for the assignment [0 0 8 0], but there are 1120 different ways of distributing the eight employees for the assignment [1 1 3 3]. There are 2520 different ways for the assignment [2 2 2 2], which is the maximum possible for our problem, but this assignment is not allowed as the total value created by this is only \$2000, which is less than the constraint \$2400.

Mathematically, as the distribution becomes broader and more uniform, there are more ways to distribute the candidates among the categories, thereby increasing the multiplicity. Therefore, by maximizing multiplicity one arrives at the broadest and the most uniform distribution that is allowed by the constraints, i.e., the fairest distribution, as we deduced above. This is exactly what the owner needs, because this distribution will give him maximum flexibility and maximum preparedness to meet an uncertain demand. This is the distribution that maximizes the robustness of the shop's operation, i.e., minimizing its downside risk, to tackle the uncertainty in the order flow.

It is easy to show from equation (4.1), that

$$\ln W = \ln N! - \sum_{i=1}^{n} \ln(Np_i)!$$

where $p_i = N_i/N$. Simplifying this by using Stirling's approximation, we get

$$\ln W = -N + N \ln N + \sum_{i=1}^{n} Np_i - \sum_{i=1}^{n} (Np_i) \ln(Np_i)!$$

$$\ln W = -N \sum_{i=1}^{n} p_i \ln p_i$$

which is essentially the same as entropy in equation (4.2).

Therefore, by maximizing entropy we are maximizing multiplicity, which means we are maximizing fairness by generating the broadest and the most uniform distribution allowed by the constraints—without assuming any other information, without rejecting any category arbitrarily (i.e. we are structurally fair), and without reducing its magnitude arbitrarily (i.e., we are parametrically fair). In a nutshell, the maximum entropy distribution treats all categories fairly.

SCENARIO 3: FAIREST INCOME DISTRIBUTION Now, let us consider a third scenario, in which the problem is kind of turned around. Suppose we are told that on a particular day, the shop sold $2400 worth of sweets. We are not given any other information besides $N = 8$ and $n = 4$. Recall that 50% of this revenue, or $1200, is to be distributed as income among the eight employees. We are asked to determine the fairest income distribution.

This requires the determination of how to distribute $1200 among the four categories. Let us assume that the team that produced this outcome was [0 0 8 0]. This certainly satisfies the constraint $V = 2400$. This assignment would then imply that all employees were to be paid equally, $150 each.

But is this a fair distribution of income? Chances are that it is not.

Why not? Even though this certainly is a possibility, for this to be true, we'd have to assume that no jobs were executed in the other three categories. What justifies this assumption? Nothing. We don't have any information that would justify this assumption. As discussed above, this assumption is arbitrary and violates the fairness principle. Therefore, it is not a fair distribution. Similarly, one can argue against all other assignments except [1 1 3 3]. Based on the information provided, this is the fairest assignment. Therefore, the fairest distribution of income is [$50 $100 $450 $600] among the four categories.

Thus, we see that the fairest treatment of the different job (i.e., value) categories, as dictated by the fairness principles, determines the fairest distribution of employees in different categories. As noted, the total value created was from the value additions of different employees working in different categories. Based on the aggregate information provided, we did not know how many employees worked in the different categories or what the different value additions were. If we had had such information, it would be straightforward to solve this credit assignment problem (i.e., who gets how much credit for creating the final value). But we don't have that information. So the fairest assignment we can make is [1 1 3 3] regarding the credit assignment. This is the fairest assignment of values contributed by the employees in the different categories toward the total value of the final products, for the given information. This, in turn, determines the fairest share of the profits among the different employees, thereby determining the fairest income distribution.

Maybe, in reality, the team that created the $2400 revenue was indeed [0 0 8 0]. In that case, the income distribution we determined would be unfair, because everyone deserved the same pay. But there is no way of knowing this reality from the information provided. The fairest decision we could make, based on the information provided, is what we found above.

There is another way of interpreting our analysis, from a statistical perspective. If Ananda Malai 108 repeated its $2400 sale thousands of times (the more, the merrier from a statistical perspective), we'd surely find the combination [0 0 8 0] occurring a few times, [2 1 0 5] a few more, [2 0 2 4] even more, but [1 1 3 3] would be the dominant occurrence. In other words, if there are thousands of replicas of Ananda Malai 108, i.e., a statistical ensemble, and if we were to analyze the $2400 sales patterns, we'd find the combination [1 1 3 3] to be overwhelmingly dominant. The state with the maximum multiplicity is the most likely outcome. This is the interpretation adopted in statistical mechanics. This leads to the superficial interpretation of entropy as randomness and disorder.

While this interpretation is acceptable for analyzing molecular behavior, it is not adequate or appropriate for analyzing the behavior of goal-driven, fairness-seeking agents such as employees. Here, we need the deeper understanding of entropy as a measure of fairness in a distribution. Indeed, one could justify the [1 1 3 3] assignment on the basis that it is the most probable outcome. But as our analysis shows, the justification is much deeper than that. It actually is the fairest distribution.

Note that we deliberately considered a simple example, so that we could work out the combinatorics to see the underlying patterns. Our theory is not meant for such small companies. It is meant only for very large corporations with tens of thousands of employees, such as the Fortune 500 companies. For such companies, the combinatorics would favor the fairest outcome overwhelmingly, as it did in the case of the equilibrium distribution of molecules in chapter 4.

Another feature favoring our approach for large corporations is that it is hard, if not impossible, to solve, even empirically, the credit assignment problem when there are tens to hundreds of thousands of employees. How does one measure reliably to determine how much value is added by different employees in different categories in such companies? In general, one cannot, even though there are exceptions, as we discuss later. We examine the general situation in section 5.5.3 and revisit this point in section 7.2 when we discuss executive compensation.

### 5.5.3  Why Can't Some Job Categories Be Undervalued?

Let us analyze one more scenario, which paves the way for the next step. Let us say that the shop makes an assortment of sweets worth $2400 every day, but only the $300 variety sells, and the others don't sell and wind up as waste. In addition, there is greater demand for $300, worth $2400 total, which the shop is not able to exploit because of its current configuration of employees. After observing this pattern for a number of days, the owner adapts by assigning everyone to category 3, and he gets rid of the other categories. This new arrangement works fine now.

Thus, we see that initially the demand pattern was stochastic, with most orders requiring an assortment of products. However, for whatever reason, the market seems to have changed, and now the demand is for only the $300 product, to which the shop has adapted. Previously, the information the owner had about the demand required the shop to make a variety of products. But now the information requires him to make only the $300 variety. Thus, when a company is successful, its internal structure and operation would closely reflect the market structure and its behavior. We call this the *lock-and-key market-company* relationship. Just as the structure of a key reveals important information about the lock, and vice versa, a company's structure and operation reveal key properties about the market, and vice versa.

This leads us to recognize and emphasize a critical property of the job categories in any company. A company exists to make money for its owners and shareholders by pursuing a business opportunity. To succeed, it needs to execute a variety of tasks efficiently. And to do this, it hires a number of employees in various job categories, who execute these tasks and create value, as the sweet shop did. Therefore, in a successful company every job category, and every employee, exists for a reason, because every job is there to address a particular business need and value creation. Otherwise, the category would not exist, as it would be a waste of resources. As noted, once our sweet shop specialized to selling only category 3 sweets, there was no need to invest in category 1, category 2, or category 4 confections. Those categories simply didn't exist anymore. This discipline is enforced by the competitive free market. If a company has employees or categories that add very little or no value to its products and services, this wasteful and inefficient operation will eventually cause the company to fail in a competitive free-market environment. Similarly, if the company does not have enough employees or job categories, again it will fail. Thus, a successful company in a free-market environment must be doing all the right things, more or less.

Therefore, the very fact that some job categories exist tells us that they address a business need and add value. Hence, unless otherwise explicitly warranted by some information, we can't reject any category or reduce its influence (i.e., magnitude) arbitrarily. *This is the fundamental reason for requiring the fairest assignment in teleological systems.* Only the fairest distribution properly acknowledges and accounts for the value added by all the employees in all the categories in a company. This is why the fairest distribution is the best or optimal distribution from a business perspective, as it maximizes the company's robustness to an uncertain demand.

### 5.5.4 The Desert View of Fairness

Our game-theoretic analysis addressed the question of fairness from the "bottom-up," *free-market dynamics* perspective. In this perspective, one asks what income distribution is produced operationally by the dynamics of a free market under ideal conditions. There is another approach based on the notion of desert in economics and philosophy (Feldman and Skow 2015). *Desert* refers to what one deserves to earn as a result of one's contributions in creating value for a company or a society, i.e., giving each person her due. This is essentially the *proportionality principle*—those who contribute more should earn more. Given that different employees make different contributions in a company, we ask, in the desert perspective, what income distribution are the employees, as a group, entitled to and deserve under ideal conditions? This is the "top-down" design perspective. What they deserve, of course, is the fairest distribution, where all three principles of fairness are satisfied. It turns out that to answer this question requires an information-theoretic approach, along the lines of the analysis of the sweet shop.

Let us revisit the large corporation iAvatars, which employs tens of thousands of employees ($N$), with a salary budget of $\$M$. All employees, ranging from low-level assistants to the CEO, contribute to the company's overall success and value creation in different ways. We build on our analysis of Ananda Malai 108, but generalize the discrete value categories into a continuous one. The continuously increasing salary scale reflects the increasing contributions in value from different employees at various levels in the corporation. We are interested in determining the fairest income distribution from this top-down perspective.

Building on our analysis of the sweet shop, we are keenly aware that all kinds of biases can easily creep in, and so we ask the following question: What do we know for certain about this distribution that is reliable and *unbiased*, a priori? There are many things we do not know about this distribution, but we do have some partial knowledge. Typically, the compensation committee in any corporation does know at least five things: (1) the total number of employees is $N$ including the CEO; (2) the total amount of money budgeted to pay all these employees is $M$; (3) the minimum salary received by the lowest employee is $S_{min}$, often fixed by the minimum-wage law; (4) the maximum salary $S_{max}$ cannot exceed $M$; and (5) different employees contribute different amounts and hence should be paid different salaries. This may not seem like much, but even this partial knowledge can help us a great deal by narrowing the choices, as we shall show, since the distribution is constrained by this information.

Any other information, besides these facts, is likely to be suspect, biased in general. For instance, all we know for sure is that different employees (or at least, different employee categories) make different contributions to the final value created by the company. But no one knows precisely how much more, for example, Bob contributed than Carol, Ted, or Alice did. These things are hard to quantify in general; at best, one has only semiquantitative assessments and educated guesses, despite the use of various performance metrics. This gets even trickier for senior management. In most cases, no one knows for sure to what degree the CEO was responsible for a company's success. There are a few exceptions, of course, where one can point to someone like Steve Jobs for the turnaround in Apple, but this is not typical. A company succeeds because a lot of talented employees work hard and execute a smart strategy well. No one knows for sure how much credit for this success should be assigned to the CEO or the senior management team (executive vice presidents, chief financial officer, chief operations officer, and so on).

Thus, beyond the basic facts listed above, we really don't have reliable additional information, in general, to make an unbiased determination. So what can we say about the distribution, given these basic facts? The question before us now is: What is the *least biased* distribution of $M$ dollars among $N$ employees as salaries $S$?

Now, let us take a step back and answer this question for the simplest case. Say that all we are given is the information that $M$ dollars are to be distributed among $N$ employees. We are not given any additional information, such as different employees contributing different amounts to the overall value of the product. Therefore, under the information provided, the fairest distribution of salaries is $M/N$, i.e., with everyone making the same salary. This is obvious as the fairest assignment because we have no information telling us that some people should be paid more and others less (as opposed to item [5] above). In the absence of such information, the fairest assignment is to treat all of them equally.

Next we are going to answer this question given that we have additional information, as listed above.

From the basic facts given above, we can derive certain useful information about the distribution. Since we know $S_{max}$ and $S_{min}$, we can write:

$$R = \ln S_{max} - \ln S_{min} = \ln M - \ln S_{min} \quad (5.7)$$

We express the range $R$ in terms of $\ln S$, and not $S$ per se, because the relevant variable is $\ln S$ and not $S$ itself, due to the diminishing marginal utility of salary. This is another piece of information we already know about the distribution. The key feature of the information-theoretic approach is to use all the known relevant information, *but nothing else*. We know that only $\ln S$ matters, not $S$ per se.

From equation (5.7), we can estimate the standard deviation $\sigma$ of the distribution by using the well-known Chebychev inequality given by (Hozo et al. 2005; Hogg and Craig 1995; Mood et al. 1974)

$$P[-a\sigma < x - \theta < a\sigma] \geq 1 - \frac{1}{a^2} \quad (5.8)$$

where $\theta$ is the mean of the distribution. By choosing a large enough value for $a$ (say, $a = 10$), we can estimate $\sigma$ as $R/2a$ with as much confidence as we desire (e.g., for $a = 10$, $P \geq 0.99$). As we shall see, the actual value of $a$ is not important for our purposes. Thus, we now know one more thing about this distribution—an estimate of its $\sigma$. Of course, we realize that this is an overestimate since the upper bound $M$ for salary is not the usual case. In practice, we would have a much more realistic upper bound. Fortunately, as we shall see, this does not affect our main conclusions.

Now, we can also estimate the mean or the expected value $E[\ln S]$, by using another well-known result, namely, Jensen's inequality (Hogg and Craig 1995; Mood et al. 1974), given by

$$E[\ln S] \leq \ln(E[S]) \quad (5.9)$$

$$\leq \ln\left(\frac{M}{N}\right) \quad (5.10)$$

which gives us an upper bound for the mean:

$$E[\ln S] = \ln\left(\frac{M}{N}\right) \equiv \mu \quad (5.11)$$

Now that we have gained additional knowledge about this distribution, let us analyze the model derived from these upper bounds.

Our central question has thus been reduced to this: What is the least biased (i.e., fairest) distribution of $M$ dollars among $N$ employees given the constraints $E[\ln S] = \mu$ and $E[(\ln S)^2] = \sigma^2$?

Again, this question is answered by applying the principle of maximum entropy, but this time from an information-theoretic perspective (Jaynes 1957a, b; Kapur 1989). The principle of maximum entropy states that given some partial information about a random variate $S$, of all the distributions that are consistent with the given information, the *least biased distribution* has maximum entropy associated with it. This is so because the maximum entropy distribution does not make any unwarranted assumptions or biases about individual values that are not explicitly specified a priori as constraints, as we discussed in the case of the sweet shop.

Following our method of Lagrange multipliers, we rediscover, once again, that the least biased distribution at equilibrium is *lognormal*, given by

$$f(S; \mu, \sigma) = \frac{1}{S\sigma\sqrt{2\pi}} \exp\left[-\frac{(\ln S - \mu)^2}{2\sigma^2}\right] \quad (5.12)$$

where $\mu = \ln(M/N) - \sigma^2/2$, $\sigma = (\ln M - \ln S_{\min})/2a$, and $a$ is a parameter chosen using the Chebychev inequality, given by

$$\text{Prob}(-a\sigma < X - \mu < a\sigma) \geq 1 - \frac{1}{a^2} \quad (5.13)$$

to the level of confidence desired in the estimate for $\sigma$ (e.g., for $a = 10$, $P \geq 0.99$). Equation (5.12) is same as equation (5.6) or (5.25) with the following identities:

$$\mu = \frac{\alpha + \gamma}{2\beta}$$

$$\sigma = \left(\frac{\gamma}{2\beta}\right)^{1/2} \quad (5.14)$$

We do realize that our estimates of the mean and standard deviation are upper bounds and that the true mean i.e., $E[\ln S]$) $\mu^*$ and standard deviation i.e., $E[(\ln S)^2]$) $\sigma^*$ of the population will be lower. Let us assume, for the sake of argument, that through additional information we are given a better estimate of the mean $\mu'$ and standard deviation $\sigma'$ which are a lot closer to the true values. Under these conditions, the principle of maximum entropy would still yield a lognormal distribution, but with a different mean $\mu'$ and standard deviation $\sigma'$. Thus, the crucial insight here is that even if our guesses about the mean and standard deviation are not very good, the *essential qualitative character of the distribution does not change and remains lognormal*—only the parameters such as the mean and standard deviation change. Furthermore, it turns out that our initial estimate for the mean as $\mu$ is not too bad. Since the resulting distribution turned out to be lognormal, the approximation

$$E[\ln S] = \ln[E(S)] \qquad (5.15)$$

is actually quite good.

In this chapter, we analyzed the free-market dynamics perspective as well as the desert perspective of the fairness question. Both perspectives lead to the same result of a lognormal pay distribution at equilibrium. It is not surprising that these two results agree as the two approaches are related perspectives on the same problem. In the desert view, we approached the problem from the information-theoretic perspective while in the free-market view we did so from the statistical thermodynamic game-theoretic perspective. In systems engineering parlance, the former is known as the *design* perspective (a top-down approach), while the latter is the *operational* perspective (a bottom-up approach) of the same system. As discussed above, in both cases we are maximizing fairness under the same constraints leading, naturally, to the same outcome.

## 5.6 Entropy: Correcting the Misinterpretations

As discussed in section 4.3, entropy is maximized at statistical equilibrium. We also saw that the equivalent defining criterion for statistical equilibrium is the equality of all accessible cells. In other words, a given molecule is equally likely to be found in any one of the accessible physical space cells (or phase space cells, in the general case) at any given time. In figure 4.2, once the partition is removed, any molecule is equally likely to be found in either chamber, as both are then accessible. When the partition was in place, as in figure 4.1, the right chamber was not accessible and hence the molecules couldn't go there.

Once the partition is removed, all unit cells, in both the chambers, are equally accessible and the molecules do not discriminate some over the others. To say that a molecule, in the long run, is more likely to be found in the left chamber than the right and to assign it a higher probability, say, $p(\text{left}) = 0.6$ and $p(\text{right}) = 0.4$ would be a biased and an unfair assignment of probabilities. This would violate the arbitrariness avoidance principle. This assignment presumes some information that has not been given. What do we know about the molecule, or its surrounding space, that would make us prefer the left chamber over the right? There is no basis for this preference. The unequal assignment of probabilities is thus arbitrary and unwarranted. Therefore, the fairest assignment of probabilities is one of equality of outcomes, i.e., $p(\text{left}) = 0.5$ and $p(\text{right}) = 0.5$. Therefore, by maximizing entropy we are, in fact, mathematically enforcing and guaranteeing the requirement that all accessible phase space cells be treated equally. In other words, this is the fairest treatment of all the accessible phase space cells.

As we discussed above, we observe, again that maximizing entropy is the same as maximizing fairness in a distribution, i.e., the fairest treatment of all possible outcomes. Thus, the *true essence of entropy is fairness*, not randomness or disorder or uncertainty.

This crucial insight reconciles the two seemingly different notions about how we reach equilibrium in our gedankenexperiment: is it by maximizing fairness in the distribution of salary (as the game-theoretic framework demonstrated in chapter 3 ) or by maximizing entropy (as the statistical thermodynamics approach showed in chapter 4)? Well, as it turns out, there is no contradiction here as they are indeed one and the same!

This critical insight about entropy as fairness has not been explicitly recognized and emphasized in prior work in statistical mechanics, information theory, or economics (Venkatasubramanian 2010). Despite several attempts in the past (Georgescu-Roegen 1987; Proops and Safonov 2004; Baumgartner 2005), entropy has played, by and large, only a marginal role in economics, even that with strong objections from leading practitioners. Its pivotal role in economics and in free-market dynamics has never been recognized in all these years. This is so mainly because entropy's essence as fairness is masked under different facets in different contexts (Venkatasubramanian 2010).

As we discussed in chapter 4, entropy is a tricky concept that has been often misunderstood, even maligned, in its 150 years of existence. It is a historical accident that the concept of entropy was first discovered in the context of thermodynamics and, therefore, it has generally been identified as a measure of randomness or disorder. Boltzmann seems to have started this when he introduced the notion of a greater "disorder" of a gas at high temperature compared to its distribution of velocities at a lower temperature to describe its higher entropy intuitively (Boltzmann 1964). Others followed suit: von Helmholtz (1883), called entropy "Unordnung" (disorder) (von Helmholtz 1883), and Gibbs described entropy, somewhat informally, as "mixed-up-ness" (Gibbs 1928; Lambert 2002). All this unfortunately has resulted in entropy getting tainted with the negative notions of doom and gloom.

Even its subsequent "rediscovery" by Claude Shannon (1948) in the context of information theory did not help much, for entropy now got associated with uncertainty or ignorance, again not a good thing. For instance, the *Oxford English Dictionary* (OED) defines entropy as "lack of order or predictability; gradual decline into disorder: e.g., 'a marketplace where entropy reigns supreme.'"

Such definitions only reaffirm the common interpretation of entropy as disorder and uncertainty. It is this state of affairs that prompted the great mathematician John von Neumann to famously advise Claude Shannon, when Shannon sought von Neumann's advice in naming the information-theoretic uncertainty measure he had discovered (Tribus and McIrvine 1971, 180): "You should call it entropy, for two reasons. In the first place your uncertainty function has been used in statistical mechanics under that name, so it already has a name. In the second place, and more important, nobody knows what entropy really is, so in a debate you will always have the advantage." Although it is funny, the sad truth is that this is still largely the case!

This misunderstanding of entropy has been a major conceptual stumbling block in economics, as we observed in section 1.5. The notion of entropy as disorder and, as Amartya Sen objected, its association with doom and gloom are problematic in economics. Paul Samuelson was dismissive outright, as we saw.

However, as our analysis reveals, the true essence of entropy is fairness, which appears with different masks in different contexts. In thermodynamics, as we just saw, being fair to all accessible phase space cells at equilibrium under the given constraints—i.e., assigning equal probabilities to all the allowed microstates—projects entropy as a measure of randomness or disorder (Reif 1965). This is a reasonable interpretation in this particular context, but it obscures the essential meaning of entropy as a measure of fairness.

In information theory, being fair to all messages that could potentially be transmitted in a communication channel—i.e., assigning equal probabilities to all the messages—shows entropy as a measure of *uncertainty* (Jaynes 1957a, b). Again, while this is an appropriate interpretation for this application, this, too, conceals the real nature of entropy.

In the design of teleological systems, being fair to all potential operating environments, entropy emerges as a measure of robustness, i.e., maximizing system safety or minimizing risk (Venkatasubramanian 2007). Once again, this is the right interpretation for this domain, but this also hides its true meaning.

Thus, when we inquire deeply into the nature of entropy, it reveals its true self, by lifting the veil of randomness and uncertainty, as a measure of fairness, which is a good thing, indeed a beautiful thing. It is not at all about doom and gloom that Sen objected to. The common theme across all these different contexts is the essence of entropy as a measure of fairness. This fairness property of entropy was hiding in plain sight. In a sense, people were using this property without explicitly recognizing and acknowledging it.

Even Jaynes (whose work I admire immensely, as can be inferred from my dedication of this book to him) seems to have missed this crucial insight of entropy as a measure of fairness. In his 1991 essay "How Should We Use Entropy in Economics?" (Jaynes 1991a), he correctly argues that thermodynamics is a closer analogy for economics than classical mechanics. While he suggests that one should consider using some sort of "economic entropy," he does not interpret what it might be in economic terms, nor does he develop his ideas into a full-fledged theory. He concludes his speculative essay (Jaynes himself recognized the speculative nature of his essay as he noted in the subtitle: "Some Half Baked Ideas in Need of Criticism") by stating (Jaynes 1991a, 5), "Therefore we think that a realistic implementation of this thermodynamic analogy still lies rather far in the future." While Jaynes noted, in passing, the "fairness" property of the maximum entropy principle in an earlier paper (Jaynes 1985), he did not pursue this for the rest of his career.

### 5.6.1 *Entropy: The Main Character in Free-Market Dynamics*

It is important not to confuse entropy as a concept from physics even though it was first discovered there. In other words, it is not like energy or momentum, which are physics-based concepts. Entropy really is a concept in probability and statistics, an important property of distributions, whose application has been found to be useful in physics and information theory. In this regard, it is more like variance, which is a property of distributions, a statistical property, with applications in a wide variety of domains.

However, as a result of this profound, but understandable, confusion about entropy as a physical principle, one got trapped in the popular notions of entropy as randomness, disorder, doom, or uncertainty, which have prevented people from seeing the deep and intimate connection between statistical theories of inanimate systems composed of nonrational entities (e.g., gas molecules in thermodynamics) and of animate, teleological, systems of rational agents seen in biology, economics, and sociology. This confusion presented in the past an insurmountable obstacle in neoclassical economics as Philip Mirowski (1989, 389) observes: "As Bausor (1987, 9) put it, whereas classical thermodynamics 'approaches entropic disintegration, [neoclassical theory] assembles an efficient socially coherent organization. Whereas one articulates decay, the other finds constructive advance.'"

This obstacle goes away once we realize that entropy is a measure of fairness, not randomness, and that maximizing entropy produces a self-organized coherent categorization of employees (which results in the lognormal income distribution), not decay or entropic disintegration. An observation Keynes made, in a different context, in his preface to his magnum opus, *The General Theory*, seems apropos here: "The difficulty lies, not in the new ideas, but in escaping from the old ones, which ramify, for those brought up as most of us have been, into every corner of our minds" (Keynes 1935, 5).

In addition, and most crucially for economics, entropy's natural connection with, and its central role in, the supply-demand driven, self-organizing, dynamics of a free market comprising rational, utility- and profit-maximizing, animate agents has not been recognized and emphasized before. As our theory demonstrates, the ideal free market for labor promotes fairness as an *emergent* self-organized property.

Now, there are a number of metrics one could use to measure fairness in a free market, such as maximin, proportional, Gini coefficient, and entropy (Venkatasubramanian 2010; Lan et al. 2010). Of these measures, what would be the most appropriate one for a free-market environment? In other words, what measure captures the kind of fairness promoted by the ideal, self-organizing, free-market dynamics as discussed in previous sections?

Recall that in the ideal free-market environment there is no central market authority to enforce those distributive justice policies that require the presence of such an authority to compute and enforce the chosen fairness measure. In practice, such distributive actions are taken afterwards through tax and transfer policies. What we are describing here is the absence of such enforcement during the real-time dynamics of the market activity. This right away would eliminate measures such as maximin, proportional, Gini coefficient, and others that would require the presence of some market authority to actively intervene in real time and execute distributive justice.

Therefore, *only entropy* is the appropriate measure of fairness for the ideal free-market environment as it does not require the presence of a market authority to compute and enforce the fairness policy. It happens by itself, naturally, through the self-organizing dynamics, the invisible hand. None of the rational agents, be it an employee or a company, in an ideal free-market system is computing any fairness measure to promote fairness. All are concerned only about their own selfish economic interests and making self-serving decisions, and the system thus evolves in a self-organizing manner. This rules out all other measures except entropy as the obvious choice.

At this juncture, it is useful to remind ourselves of our definition of an ideal free market, as spelled out in section 3.1., to clarify the role of a central market authority. We defined our ideal free market as an environment that is perfectly competitive, where transaction costs are negligible and no externalities are present. There are no biases due to race, gender, religion, etc. We assumed that no single agent (or a small group of agents), whether an employee or a company, can significantly affect the market dynamics; i.e., there is no rent seeking or market power exploitation. In other words, our ideal free market is a level playing field for all its participants. We also assumed that neither the companies nor the employees engage in illegal practices such as fraud, collusion, and so on. We also observed that all these requirements do not happen by themselves, and that the free market, ideal or otherwise, itself is a human creation, and "does not exist in the wilds beyond the reach of civilization," as Reich (2015, 4) writes. In that sense, the ideal free market requires the presence of some authority that enforces these basic rules of the game.

When we stated above that there is no central market authority to enforce the fairness policies, what we meant was that there is no such authority to enforce any *distributive justice policy*. That is, no one is there to make sure that maximin, proportional, or Gini index–based fairness metrics are followed during market dynamics. We do, however, need an authority to enforce the rules of the game so that the resulting free market behaves ideally.

### 5.6.2 Entropy and the Invisible Hand: The Fair-Market Hypothesis

We believe that by properly recognizing entropy as a measure of fairness and demonstrating how it is naturally and intimately connected to the dynamics of the free market, our theory makes a significant conceptual advance in revealing the deep and direct connection among game theory, statistical thermodynamics, information theory, and economics.

This revelation of entropy's true meaning also sheds new light on a decades-old fundamental question in economics, as Samuelson (1972, 260) posed in his Nobel Lecture: What is it that Adam Smith's invisible hand is supposed to be maximizing? Or as Monderer and Shapley (1996, 125–26) stated regarding the potential function $P^*$ in game theory, "This raises the natural question about the economic content (or interpretation) of $P^*$: What do the firms try to jointly maximize? We do not have an answer to this question."

Our theory suggests that what all the agents in a free-market environment are jointly maximizing, i.e., the invisible hand is maximizing, is fairness. Maximizing entropy, or the game-theoretic potential, is the same as maximizing fairness in economic systems, i.e., being fair to everyone under the given constraints. In other words, economic equilibrium is reached when all agents feel they have been fairly compensated for their efforts.

As we all know, fairness is a fundamental economic principle that lies at the foundation of the free-market system. This notion of fairness, again, is founded on our three principles of fairness, particularly the equality principle. It is so vital to the proper functioning of the markets, and for a democratic society, that we have regulations and watchdog agencies that break up and punish unfair practices such as monopolies, collusion, price gouging, and insider trading, thus enforcing the aforementioned rules of the game. Thus, it is eminently reasonable, indeed particularly reassuring, to find that maximizing fairness, i.e., maximizing entropy, is the condition for achieving economic equilibrium. We call this result the *fair-market hypothesis*.

We claim that the ideal free market, in addition to being Pareto-efficient, promotes fairness to the maximum level allowed by the constraints imposed. Thus, there is an additional role of the ideal free market, the fairness objective, which is generally not acknowledged in conventional economic theories. We believe that the true value of the ideal free-market system comes from its ability to achieve both efficiency and fairness. In our thought experiment, we achieved Pareto efficiency by enforcing the constraints that both $M$ and $N$ are to be satisfied. So this is done automatically. What was left was the fairness objective. Through the self-organizing adaptive dynamics, the second objective was also met by maximizing the potential. Fairness turns out to be a critical property for the stability of a free market, which we discuss in sections 5.9 and 8.2.

A related interpretation is that the game-theoretic potential captures the trade-offs among utility from salary, disutility from effort, and utility from a fair opportunity for future prospects, for all the agents collectively. The ideal free market tries to accommodate every agent's individual preference regarding this trade-off, given the overall constraints on money and job openings. Thus, in a sense, the market is trying to maximize "harmony," an accord freely and jointly agreed to by all the agents, where all agents feel fairly compensated for their efforts.

To switch directions a bit, if we have no information about the choices, as we discussed, then a uniform distribution (i.e., equal probabilities) is the fairest assignment. What if we have additional information in the form of constraints? Generalizing the principle of insufficient reason to this case, we get the principle of maximum entropy (PME) (Jaynes 1957a, b) which answers the question; What is the fairest assignment of probabilities among several alternatives, given a set of constraints? The answer, of course, is the assignment that maximizes entropy under the given constraints. Thus, the roots of the principle of maximum entropy as the principle of maximum fairness can be traced all the way back to the principle of insufficient reason (Venkatasubramanian 2010).

In table 5.2 we list some sample maximum entropy distributions, given various pieces of information as constraints on the distribution.

TABLE 5.2
Maximum Entropy Distributions

| Constraint specified | Distribution |
|---|---|
| No prior information | Uniform |
| Mean $\langle x \rangle$ | Exponential |
| Mean $\langle x \rangle$ and variance $\langle x^2 \rangle$ | Normal |
| Geometric mean $\langle \ln x \rangle$ | Pareto (i.e., power law) |
| $\langle \ln x \rangle$ and $\langle (\ln x)^2 \rangle$ | Lognormal |

There are also certain other conditions, such as whether the variate is discrete or continuous, its domain, etc. For these details, the reader is referred to Kapur (1989).

TWO RELATED NOTIONS OF FAIRNESS  It is useful to clarify the two different, but related, uses of the term *fairness* in our discussion.[2] One is in the Nozickian perspective, and the other is the information-theoretic perspective.

According to Nozick, any distribution that arises *naturally* as a result of the free-market interactions is fair, by definition. For Nozick, this outcome is fair because it was the result of legitimate transactions, jointly and voluntarily agreed to, among consenting individuals. He considered it fair because it was arrived at freely, without any coercion, by its participants. In claiming this, Nozick is assuming implicitly an ideal free market with the properties and restrictions we outlined in section 3.1.

In the information-theoretic perspective, the fairest outcome is realized when one maximizes entropy.

To illustrate and clarify this point, consider a group of $N$ people, among whom we have to distribute $100,000 of resources. Since there is no reason to prefer one person over another, the fairest distribution is obvious—each person gets $100,000/$N$.

Now, one could easily make the mistake of thinking this is not the maximum entropy distribution. One might erroneously conclude the maximum entropy outcome will be an exponential distribution (for the case of a variate, i.e., the payoff here, that varies continuously in the semi-infinite interval $[0, \infty)$ (Kapur 1989)), with a majority of people getting almost nothing and a few getting a lot.

However, each person getting $100,000/N$ is, in fact, the maximum entropy outcome. There is a subtle point here that is easy to miss.

The most general formulation of this problem addresses how to distribute $N$ people among $n$ different categories of payoffs such that the total payoff is $\$M$ (which is $100,000 in our example). Given $N$, $n$, and $M$, the fairest distribution is found by maximizing entropy given by equation (4.2) ($k_B$ is 1 in our example). Under these conditions, the fairest distribution is indeed exponential.

However, here is the subtle point. The exponential outcome is valid only when it is already known that there are $n$ different categories of payoffs. That is, the exponential outcome presumes, or it is given, that there are different categories of payoffs. This presumption, or stated explicitly a priori, assumes that the recipients are all different. Therefore, some will receive more funds, and others less, thereby creating different categories of payoffs.

But when all recipients are treated as equals (i.e., it is not explicitly stated that there are different classes of recipients), then we are required to treat them all as equals. This is a consequence of the principle of arbitrariness avoidance, which is a restatement of the principle of insufficient reason. This is generalized to the principle of maximum entropy as we saw above. Thus there cannot be different categories of payoffs. We cannot assume there are different categories of payoffs because (1) we have not been told so explicitly and (2) we have not been told that the recipients are also different from one another in such a way that different recipients require different compensation.

Since there are no different categories of payoffs, we must assume there is only one category. Therefore, all $N$ recipients belong to this category of payoff. That is, all $N$ get the *same* payoff, which, of course, is $100,000/N$.

Therefore, this is simply the result of maximum entropy, except that in this case the principle of maximum entropy reduces to its simplest form, namely, the principle of insufficient reason. As we proved above, the ideal free market maximizes fairness in income distribution— this is what we call the *fair-market hypothesis*. This is the claim Nozick makes implicitly, as noted above. Therefore, these two notions of fairness are indeed related. To Nozick, it is justified based on philosophical grounds. To us, it is justified based on information-theoretic arguments.

## 5.7 Replicator Dynamics

We now develop yet another approach, by bringing another powerful tool from the arsenal of game theory, namely, the method of *evolutionary dynamics*. Evolutionary game theory is another approach to studying the behavior of large populations of agents who repeatedly engage in strategic interactions. Its origins can be traced to the work of John Maynard Smith in the context of evolutionary biology, which studies the emergent dynamic behavior and characteristics of large populations of organisms (Smith 1972, 1974, 1982). Evolutionary biology is based on the concept that an organism's genetic makeup largely determines its properties and therefore its fitness for survival and growth in a given environment. Fitter organisms will tend to produce more offspring, thereby enabling genes that cause greater fitness to increase their presence in the population. Over time, fitter genes come to dominate the population because they lead to higher rates of reproduction.

As an organism competes for the limited resources in a given environment for its survival and growth, its success depends on the behaviors it displays in garnering resources in comparison with those of the competing organisms. To see this from the perspective of game theory, an organism's genetically hardwired features and behaviors are its strategies, its fitness payoff, which depends on the strategies of the competing organisms. An important modeling framework that is commonly used to address such evolutionary dynamics problems is the *replicator dynamics* approach (Sandholm 2010; Hofbauer and Sigmund 1998).

In this framework, one models how a fraction of a population adopting a particular strategy evolves over time, to determine the population's equilibrium properties. Let us assume that a population of organisms is divided into $n$ types of species $\Upsilon_1$ to $\Upsilon_n$ which are present in proportions $x_1$ to $x_n$, respectively, in the population. The fitness $h_i(\mathbf{x})$ of $\Upsilon_i$ will be a function of the composition of the population, given by $\mathbf{x}$. The rate of increase of $\dot{x}_i/x_i$ of type $\Upsilon_i$ is a measure of its evolutionary success. This is proportional to the extra fitness enjoyed by the species $\Upsilon_i$ over the rest of the population, i.e., fitness of $\Upsilon_i$ ($h_i$)– average fitness ($\bar{h}$). This results in the replicator equation

$$\dot{x}_i = x_i(h_i - \bar{h})$$

In this approach, an employee revises her strategy based on

$$\rho_{ij} = x_j[h_j - h_i]_+ \tag{5.16}$$

Under this protocol, an employee in job category $i$ who receives a revision opportunity, i.e., a new job offer in category $j$, switches from $i$ to $j$ with probability $\rho_{ij}$.

Therefore the dynamic becomes

$$\dot{x}_i = \sum_{j=1}^{n} x_j \rho_{ji} - x_i \sum_{j=1}^{n} \rho_{ij} \tag{5.17}$$

$$\dot{x}_i = x_i \left( \sum_{j=1}^{n} x_j[h_i - h_j]_+ - \sum_{j=1}^{n} x_j[h_j - h_i]_+ \right) \tag{5.18}$$

Define two sets $A$ and $B$ such that

$$A = \{k | h_k \leq h_i\}$$
$$B = \{k | h_k > h_i\}$$

$$\dot{x}_i = x_i \left( \sum_{j \in A}^{n} x_j(h_i - h_j) - \sum_{j \in B}^{n} x_j(h_j - h_i) \right) \tag{5.19}$$

This reduces to

$$\dot{x}_i = x_i \left( h_i - \sum_{j=1}^{n} x_j h_j \right) \tag{5.20}$$

The equilibrium is reached (i.e., $\dot{x} = 0$) when the individual payoff equals the average payoff of the system (we ignore the trivial solution of $x_i = 0$):

$$h_i^* = \sum_{j=1}^{n} x_j h_j^* = h^* \tag{5.21}$$

$$h^* = \alpha \ln S_i - \beta (\ln S_i)^2 - \gamma \ln N_i \tag{5.22}$$

$$N_i = \exp\left[\frac{\alpha \ln S_i - \beta (\ln S_i)^2 - h^*}{\gamma}\right] \tag{5.23}$$

$$x_i = \frac{1}{N} \exp\left[\frac{\alpha \ln S_i - \beta (\ln S_i)^2 - h^*}{\gamma}\right] \tag{5.24}$$

Rearranging equation (5.24), we find the equilibrium distribution to be

$$x_i = \frac{1}{S_i D} \exp\left\{-\frac{[\ln S_i - (\alpha + \gamma)/2\beta]^2}{\gamma/\beta}\right\} \qquad (5.25)$$

where $D = N \exp\left[h^*/\gamma - (\alpha + \gamma)^2/4\beta\gamma\right]$. This result agrees with equation (5.6). The average payoff at equilibrium is given by

$$h^* = \gamma \ln Z - \gamma \ln N \qquad (5.26)$$

where $Z = \sum_{j=1}^{n} \exp\{[\alpha \ln S_j - \beta(\ln S_j)^2]/\gamma\}$ resembles the partition function seen in statistical mechanics.

For the thermodynamic game, it is easy to show from equations (5.16) through (5.21), and (5.1), that a similar replicator dynamics analysis produces the same Gibbs-Boltzmann exponential distribution in equation (5.4) at equilibrium.

The intuitive connection between the potential game formulation and the replicator dynamics approach is the following: In the former, the potential keeps track of the total utility change in the population as the agents switch jobs, whereas in the latter the utility change is tracked one agent at a time. The end result is the same at equilibrium.

Thus, we see that, maximizing the game-theoretic potential [equation (3.11) or (5.2)] is the same as maximizing entropy subject to the constraints. In the statistical mechanics or information-theoretic formulations, these constraints are separately imposed on entropy whereas in the game-theoretic formulation [equation (3.11) or (5.2)] the constraints are already embedded in the equation (the only additional constraint imposed is the total number of agents $N$). Therefore, the resulting Lagrangian [e.g., equation (5.3)] is the same, thereby leading to the same distribution. These demonstrate the internal consistency among the four different approaches, namely, potential game theory, replicator dynamics, statistical mechanics, and information theory, which is reassuring.

As noted, one could presumably choose other expressions to model the three elements in equation (3.1), but it is not clear whether they will necessarily lead to the Gibbs-Boltzmann distribution, Helmholtz free energy, and entropy in the limiting case of the thermodynamic game involving molecules. We find this correspondence to be particularly appealing, in fact comforting, that statistical teleodynamics properly reduces to well-known results in statistical thermodynamics as a limiting case. This universality has a nice ring to it.

Now that we have answered the first set of two questions, we turn our attention toward the second set of two questions regarding the morality and stability of a free-market economy.

## 5.8  Is the Ideal Free Market Morally Justified?

As discussed in sections 1.4 and 1.5, morality is an elusive, broad concept as it can mean different things to different people in different contexts. For instance, Michael Sandel provides an introductory overview of the complexities and subtleties involved in the different notions of morality and justice in his excellent book *Justice* (Sandel 2009). However, in our work we consider morality only in a narrow sense of the term that is relevant to distributive justice. In its essence, morality is about discriminating between right and wrong. In the context of distributive justice, which is the focus of our theory, many believe that it is wrong to have a large majority of the working population struggling while a small minority is enjoying the fruits that the majority helped produce. To recall Scanlon (2005): "But the objections to inequality that I have listed rest on a different moral relation. It's the relation between individuals who are participants in a cooperative scheme. Those who are related to us in this way matter morally in a further sense: they are fellow participants to whom the terms of our cooperation must be justifiable. In our current environment of growing inequality, can such a justification be given?"

So here morality is simply about whether the free-market system is benefiting a majority of the people or only a small minority. Reich (2015) and Stiglitz (2011) also seem to interpret morality in this way. So in our theory, we follow this interpretation of morality. This is a more restrictive sense of the term, and we do not concern ourselves with its broader connotations, as discussed by Sandel (2009).

Some have claimed, however, that the free market is neither moral nor immoral, but *amoral*. For instance, Paul Krugman, Nobel Laureate in Economics, argues as follows in his essay "Economics Is not a Morality Play" (Krugman 2010): "But maybe this is an opportunity to reiterate a point I try to make now and then: economics is not a morality play. It's not a happy story in which virtue is rewarded and vice punished. The market economy is a system for organizing activity—a pretty good system most of the time, though not always—with no special moral significance."

It is even argued that Adam Smith's *The Wealth of Nations* itself is an amoral theory of the free-market economy because Smith does not make overtly moralistic arguments defending it; see, e.g., Brown (1994). However, Samuel Fleischacker (2004a, 49) disagrees:

> *The Wealth of Nations* is shot through with remarks like the following.
> > It is but equity... that they who feed, cloath, and lodge the whole body of the people, should have such a share of the produce of their own labor as to be themselves tolerably well fed, cloathed, and lodged.
> >
> > All for ourselves, and none for other people, seems, in every age of the world, to have been the vile maxim of the masters of mankind.
>
> These remarks are not morally neutral, and they presuppose that the agents they criticize are capable of living up to norms of equity, humanity, generosity, and justice.

While Fleischaker agrees that the "moral concerns are muted in *The Wealth of Nations*," he goes on to argue the main reasons for Smith's moral approval of capitalism. Thus, it appears that Smith also was concerned about the moral justification of the free market, like Scanlon, Reich, and Stiglitz, in terms of benefiting the economic well-being of the majority of the people.

As we all know, Smith claimed that by letting all individuals pursue their selfish drive to maximize their utility, the society also will benefit as an emergent phenomenon, even though it was not originally intended by the individual agents, through the action of the invisible hand. This is the intuition behind his famous remark: "It is not from the benevolence of the butcher, the brewer, or the baker that we expect our dinner, but from their regard to their own interest. We address ourselves not to their humanity but to their self-love, and never talk to them of our own necessities, but of their advantages" (Smith 1976, 27).

In other words, Smith and his followers claimed that the ultimate outcome is beneficial to the entire society, increasing the total utility of the population. Free-market advocates point to the quality of life improvements made in their societies for hundreds of millions of people over the last two centuries. This beneficial outcome, in their view, provides the moral justification.

In addition, beyond the argument that "the ends justify the means," Schoenberg (2011) observes that free-market fundamentalists believe in the intrinsic morality of free markets: "Indeed, I think that widespread support for free markets is based more on belief in their inherent morality than on belief that they promote economic growth, potentially explaining the religious fervor of free-market fundamentalists defending their faith despite the considerable counter-evidence provided by recent events." We recall that Nozick defended his free market–based libertarianism as inherently fair and moral.

Historically, all discussions on the moral foundations of the free market, both for and against, have generally been based on either qualitative arguments or empirical observations, or a combination of both, such as the ones above. However, it would be quite valuable if we could answer this question of free-market morality, in the context of distributive justice, one way or the other, using a mathematical framework, at least under ideal conditions. This is what we accomplish next, in the form of two important results.

### 5.8.1 Equality of Welfare at Equilibrium

As we note from equation (5.26), when income is distributed lognormally at equilibrium, every agent in BhuVai is enjoying the *same level of effective utility* (i.e., the same level of happiness or welfare) $h^*$, derived from the income received for one's contributions. Equation (5.26) can also be written as

$$h^* = \gamma \ln \left( \frac{Z}{N} \right) \qquad (5.27)$$

Put simply, everyone is *equally happy* at the level of $h^*$. Everyone is *not* making the same salary, of course, but they are all equally happy. Recall that Reich, Scanlon, and Stiglitz (and others) asked why should anyone voluntarily participate in a free market enterprise if the benefits go mostly to a select few? Now we have the reason, the *moral justification*, because *no one is left behind* in an ideal free market. Not only does *everyone* benefit but also everyone benefits to the *same extent*. This is the reason for everyone to participate voluntarily in a free-market capitalist economy, under ideal conditions.

## 5.8.2 Social Optimality at Equilibrium

Now we prove another important result that also justifies the moral foundations of an ideal free-market economy. Here we ask the question: Is the ideal free market socially optimal?

A *socially optimal* (SO) distribution is one in which the overall effective utility of the entire population (i.e., the *total effective* utility of the society) is maximized. This is the outcome desired by utilitarians such as Jeremy Bentham and John Stuart Mill.

To answer this question, let us restate equation (3.6) as

$$h_i = \gamma(R_i - \ln N_i) \tag{5.28}$$

where $R_i = [\alpha \ln S_i - \beta(\ln S_i)^2]/\gamma$.

The Nash equilibrium (NE) distribution (i.e., lognormal) may now be written as

$$N_i^{NE} = N\left\{\frac{e^{R_i}}{Z}\right\} \tag{5.29}$$

where $Z = \sum_{i=1}^{n} e^{R_i}$ is the partition function introduced before.

Now let us maximize total effective utility instead of potential. The Lagrangian is given by

$$L = \sum_{i=1}^{n} N_i h_i + \lambda \left(N - \sum_{i=1}^{n} N_i\right) \tag{5.30}$$

Solving $\partial L/\partial N_i = 0$, we obtain

$$\gamma R_i - \gamma \ln N_i - \gamma - \lambda = 0 \tag{5.31}$$

Therefore, the socially optimal distribution is

$$N_i^{SO} = e^{R_i - 1 - \lambda/\gamma} \tag{5.32}$$

Applying the normalization condition that $\sum_{i=1}^{n} N_i = N$, we obtain

$$e^{1+\lambda/\gamma} = \frac{1}{N}\sum_{i=1}^{n} e^{R_i} \tag{5.33}$$

Therefore, equation (5.32) reduces to

$$N_i^{SO} = N\left\{\frac{e^{R_i}}{Z}\right\} \tag{5.34}$$

which is the same as the Nash equilibrium income distribution given by equation (5.29).[3]

This is a valuable result for it proves that by maximizing the potential (i.e., fairness) one is indeed maximizing the total effective utility of the entire population, which is a very desirable outcome. Again, recall that Smith claimed that by letting all individuals pursue their selfish drive to maximize their utility, the society also will benefit, through the action of the invisible hand. This result is a mathematical proof of his claim.

At first glance, this result could be surprising because Nash equilibrium outcomes are *not* generally socially optimal. However, upon deeper reflection, the equality of the two distributions should not be surprising. By maximizing the potential to reach Nash equilibrium, we were maximizing fairness, as we discussed above. This means all citizens have been treated fairly, receiving their fair share of the income pie in exchange for their contribution. This is the best possible outcome that a society can hope for. When everyone is treated fairly, it naturally follows that the society automatically benefits the most. In other words, when everyone is treated fairly, i.e., when commutative justice is achieved at the individual level for all individuals, it automatically results in distributive justice at the system level, i.e., at the company or the societal level. Treating everyone fairly is the foundational principle of organizing a free market–based capitalist society. This automatically leads to maximizing the total effective utility of the society.

The social optimality automatically implies that this outcome is also *Pareto-optimal*. If this outcome weren't Pareto-optimal, this implies that there would exist a different outcome in which all individual utilities were at least as large, and one larger. Therefore, this would be an outcome with a *larger* total utility, which would contradict the social optimality result we just proved. Hence, the socially optimal outcome is also Pareto-optimal.

Thus, we have proved mathematically, in two different but related ways, that the invisible hand is indeed doing what it is supposed to be doing, in an ideal free market for labor, in the context of distributive justice. Therefore, the free market makes sense, under ideal conditions, not only from the perspective of economic *efficiency* but also *morally*. The ideal free market is intrinsically moral, as far as distributive justice is concerned. Adam Smith's trust in the ideal free market was not misplaced after all!

But real-life free markets are a different story, as we all know too well. Real-life free markets are not level playing grounds, where rent seeking is rampant, market power is constantly exploited, and the influence of the working class is steadily waning. When a free market is not functioning ideally, the overall utility enjoyed by the society as a whole goes down, thereby weakening the moral underpinnings of the social contract that is at the very foundation of a free-market economy, as Reich, Scanlon, and Stiglitz have argued.

### 5.8.3  Why Are NE and SO Distributions the Same?

Despite the intuitive insight we provided above, one might still wonder under what mathematical conditions these two outcomes are the same. We pursued this line of inquiry and provide below the mathematical insight for this curious result.

Let us define

$$h_i = \gamma[\,R_i + w_i(N_i)\,] \tag{5.35}$$

where $R_i = [\alpha \ln S_i - \beta(\ln S_i)^2]/\gamma$ and $w_i(\cdot)$ is the utility from a fair opportunity for upward mobility, defined in section 3.2.

Nash equilibrium implies that

$$h^* = \gamma[\,R_i + w_i(N_i)\,] \tag{5.36}$$

In other words,

$$w_i(N_i) = C - R_i \tag{5.37}$$

where $C$ is a constant.

Now, social optimum implies [i.e., from equations (5.30) and (5.31)]

$$\frac{\partial L}{\partial N_i} = \gamma \left[ R_i + \frac{d}{dN_i}[N_i w_i(N_i) - \lambda] \right] = 0$$

That is,

$$\frac{d}{dN_i}[N_i w_i(N_i)] = D - R_i \tag{5.38}$$

where $D$ is a constant.

Equating $R_i$ in equations (5.37) and (5.38) and simplifying, we get

$$w'_i(N_i) \propto \frac{1}{N_i} \qquad (5.39)$$

The only function that satisfies this condition is a logarithm. Thus, for the NE and SO distributions to be the same, we must have[4]

$$w_i(N_i) \propto \ln N_i \qquad (5.40)$$

which is what we have from equation (3.2): $w_i(N_i) = -\gamma \ln N_i$.

We now address the last remaining question.

## 5.9 Is the Equilibrium Income Distribution Stable?

As Reich, Scanlon, Stiglitz, and others have argued, the fundamental immorality of the extreme inequality in income and wealth is a concern regarding societal stability. When people feel that the system is unfair, they lose faith in the system and engage in behaviors that could ultimately bring down the system. While we can appreciate this intuitively, can we address this issue mathematically? As it turns out, we can, by appealing to the concept and techniques of stability in dynamical systems. This shouldn't be surprising for, after all, the free market and the society at large are dynamical systems with millions of agents interacting.

The stability of dynamic systems has had a long and storied history, which we won't go into, featuring some of the greatest minds in mathematics and physics (Ball and Holmes 2008). It all started with questions regarding the stability of the solar system, and the control of steam engines. Simply stated, one is interested in the following question: If I were to perturb a system when it is at its equilibrium point, would it continue to stay at the equilibrium point, or would it run away from it?

For example, consider a marble sitting precariously on top of a U-shaped cup kept upside down. A little nudge to the marble will cause it to roll down on the side, under the influence of gravity, never to return to its original position. Hence, the equilibrium point is said to be *unstable*. On the other hand, if we held the cup right side up and dropped the marble into it, it would quickly settle down at the bottom. If we perturbed it, it would move about its equilibrium point for a little while and then return to its original state. This equilibrium is said to be *stable* (actually *asymptotically stable*, see below).

As another example, consider a simple pendulum, such as an iron ball hanging from the ceiling at the end of a long rope, initially at rest, at its equilibrium point. Now, if you pull the ball a little bit, away from its equilibrium point, and let go, it will begin to oscillate about its equilibrium point. If there is no frictional loss (e.g., due to air resistance), the ball will keep oscillating about the equilibrium point indefinitely. That is, it always stays in the same neighborhood of the equilibrium point for all time and does not run away from it. Thus, the equilibrium is *stable*. If there is air resistance, then the frictional losses to energy will cause the oscillations to die down, ultimately returning the iron ball to its original position of rest. This outcome is also stable, but it is called *asymptotically stable*, as it returns to the original location eventually. Thus, there are three possible outcomes when an equilibrium state is disturbed—unstable, stable, and asymptotically stable.

Mathematically, stability issues are addressed by using the techniques developed by the Russian mathematician Aleksandr Mikhailovich Lyapunov (Aström and Murray 2008). Lyapunov proved that by analyzing the properties of a characteristic scalar function associated with the dynamical system $V(x, t)$, called the *Lyapunov function*, and its time derivative along the trajectory $\dot{V}$, one can determine the stability aspects of an equilibrium.

A Lyapunov function $V$ is a continuously differentiable function that takes positive values everywhere except at the equilibrium point (so $V$ is *positive definite*), and decreases (or is nonincreasing) along every trajectory traversed by the dynamical system ($\dot{V}$ is *negative definite* or *negative semidefinite*). For instance, for physical systems such as the pendulum or the marble in a cup, the energy function is the Lyapunov function, whose value decreases along the trajectory, as time passes, due to frictional losses. Lyapunov's insight was to realize that if the system loses energy over time and the energy is not replenished, then the system must eventually come to a stop and reach some final resting state.

The stability of an equilibrium, in the sense of Lyapunov, means that solutions starting "close" to the equilibrium (say, within a distance $\delta$ from it) remain close forever (say, within a distance $\epsilon$ from it). Note that this must be true for any $\epsilon$ that one might choose. Asymptotic stability, in the sense of Lyapunov, means that solutions that start close not only remain close, but also eventually converge to the equilibrium ($x_e$).

Lyapunov functions are very useful because one can determine the stability of a dynamic system *without* actually solving the underlying dynamical equations. However, the drawback is that finding a Lyapunov function for the given system is not easy, often more of an art than a science.

Lyapunov proved that if one can find such a function, then the system is *locally stable* at equilibrium if $\dot{V}$ is *negative semidefinite* and is *asymptotically stable* and if $\dot{V}$ is *negative definite*. A system is globally asymptotically stable if every trajectory converges on $x_e$ eventually.

Now that we have highlighted the essential aspects of Lyapunov stability, we are ready to answer our last question. We identify our Lyapunov function as

$$V(\mathbf{x}) = \phi^*(\mathbf{x}) - \phi(\mathbf{x}) \qquad (5.41)$$

where $\phi^*(\mathbf{x})$ is the potential at the Nash equilibrium (remember that $\phi^*$ is at its maximum at NE) and $\phi(\mathbf{x})$ is the potential at any other state. Note that $V(\mathbf{x})$ has the desirable properties we seek: (1) $V(x_{NE}) = 0$ and $V(\mathbf{x}) > 0$ elsewhere; that is, $V(\mathbf{x})$ is *positive definite*. (2) Since $\phi(\mathbf{x})$ increases as it approaches the maximum, $V(\mathbf{x})$ decreases with time, and hence it is easy to see that $\dot{V}$ is *negative definite*. Therefore, the equilibrium is not only stable but also *asymptotically stable*.[5]

Thus, we have proved that the equilibrium income distribution is *unique, optimal, and asymptotically stable* in an ideal free market.

## 5.10 Fairness, Stability, and Robustness of a Free-Market Society

This mathematical proof validates our intuition about fairness and stability. Recall that at equilibrium, fairness is maximized. Our Lyapunov function essentially measures the *unfairness* in the distribution. As people switch jobs while seeking better utilities, as we saw in our gedankenexperiment in section 3.1, the ideal free market's self-organizing dynamics improve fairness in the distribution. Thus, unfairness decreases as the distribution evolves toward equilibrium, which is modeled by the decreasing value of the Lyapunov function. At equilibrium, unfairness is zero, since we have reached the state of maximum fairness.

Intuitively, as well from historical lessons, we know that fairness matters a great deal in the long-term stability of societies. If great wrongs persist, sooner or later there will be popular uprisings to correct the inequities. One can perpetrate unfairness and keep the people oppressed by force for only so long. History teaches us that there will soon be a day of reckoning, a Bastille Day. If inequities are not corrected through timely, feedback control style, incremental improvements, the eventual outcome will be more violent structural adjustments.

The United States has seen its share of both. The monstrous unfairness of slavery was corrected through a civil war. Slavery was "efficient" in the eyes of the plantation owners, keeping their costs down while maximizing profits. But the grotesque unfairness of this efficient organization ultimately led to its violent demise. Even though this organization was efficient, it did not survive in the long run—it was efficient but not stable. Other inequities, such as women's suffrage, were remedied through less violent means, but after a struggle nevertheless.

The dawn of the industrial age also saw similar improvements to equity to save capitalism. As Heilbroner (1999, 105–107) reminds us:

> England in the 1820s was a gloomy place to live; it had emerged triumphant from a long struggle on the Continent, but now it seemed locked in an even worse struggle at home. For it was obvious to anyone who cared to look that the burgeoning factory system was piling up a social bill of dreadful proportions and that the day of reckoning on that bill could not be deferred forever.
>
> Indeed, a recital of the conditions that prevailed in those early days of factory labor is so horrendous that it makes a modern reader's hair stand on end. In 1828, *The Lion*, a radical magazine of the times, published the incredible history of Robert Blincoe, one of eighty pauper children sent off to a factory at Lowdham. The boys and girls—they were all about ten years old—were whipped day and night, not only for the slightest fault but to stimulate their flagging industry.... At Litton the children scrambled with the pigs for the slops in a trough; they were kicked and punched and sexually abused.... But with full discount for exaggeration, the story was nonetheless all too illustrative of a social climate in which practices of the most callous inhumanity were accepted as the natural order of events and, even more important, as nobody's business. A sixteen-hour working day was not uncommon, with the working force tramping to the mills at six in the morning and trudging home at ten at night.... Southey, the poet, wrote, "At this moment nothing but the Army preserves

us from that most dreadful of all calamities, an insurrection of the poor against the rich, and how long the army may be depended upon is a question which I scarcely dare ask myself"; and Walter Scott lamented, "the country is mined beneath our feet."

Of course, and thankfully, we have come a long way from those horrible conditions in Western capitalist societies through incremental improvements (but the continued practice of "sweatshops" elsewhere is of great concern requiring further progress). For example, the Factory Act of 1819 and the Factory Act of 1833 limited child labor, and the Factory Act of 1847 limited the working hours to ten hours per day. A further series of Acts improved the working conditions and worker safety, and the Trade Boards Act of 1909 introduced the minimum wage. While these examples are from the United Kingdom (since we recounted the working conditions there), similar laws toward improving fairness were passed in other countries over the last century.

The point of this discussion is that even in a capitalist society, which is driven essentially by the profit motive, no matter how expedient efficiency is (child labor, long working hours, etc., were all efficiency-driven measures to lower costs and increase profits), if the resulting unfairness is excessive, as it was in these cases, then the system will eventually break down violently unless mid-course corrections are made in a timely manner. Fairness determines the *long-run stability and survival* of the system.

Therefore, at equilibrium, where everyone has been rewarded fairly, there is no reason for anyone to do anything to upset the capitalist system. Even though the stability proof is valid for only small perturbations,[6] one could see potentially how the system might survive larger upsets. Since people believe in the system because it is maximally fair, everyone pulls together when the system is under attack, to protect the system that has been good to them. As a result, the ideal free market has a natural tendency to be *robust* to shocks and has the *resilience* to get back to normalcy.

Now we need to be careful here about overgeneralizing from *stability* to *robustness*. Even though they seem to mean the same thing (and sometimes they do) and are often used interchangeably in common parlance, there are important technical differences and nuances (Jen 2005).[7] Generally speaking, as discussed above, stability refers to the property of deviating a small amount from the equilibrium state when a particular input is changed by a small amount. Thus, it usually applies to input and output variables that can be quantified and varied more or less continuously. However, robustness is a broader concept that refers to a much wider class of disturbances, many of which cannot be quantified by using some metric. For example, we call a software system *robust* if it can deliver its design functionality despite user errors. Another example would be an immune system that can successfully defeat a new virus threat. User errors or new viruses are hard to quantify as a continuous metric, which is what would be required to carry out the traditional stability analysis.

Nevertheless, the stability result is helpful to make some useful observations. From a systems engineering perspective, the importance of stability and robustness in free-market operation makes perfect sense as engineers design systems with two key objectives in mind: efficiency and robustness (Venkatasubramanian et al. 2004; Venkatasubramanian 2007). In fact, engineers rarely design systems with efficiency as the sole objective as we know from practical experience that our systems are always subject to uncertainties, disturbances, and shocks, all of which our systems have to survive and perform. Imagine designing a car by caring only about fuel efficiency. Such an emphasis would naturally lead to an extremely light weight automobile that might give, say, 150 miles per gallon, which is, of course, terrific. But such a vehicle is unlikely to survive even an average collision on the road. In other words, the car is *efficient* but *not robust*.

On the other hand, imagine designing a car by focusing only on robustness to collisions. This design would lead to an automobile that might look more like a tank than a car with a very low fuel efficiency. This vehicle is *robust* but *not efficient*. No car is designed with only efficiency or robustness as the objective. Generally, efficiency and robustness requirements tend to oppose each other; i.e., highly efficient systems tend not to be robust and vice versa. So the design would typically require a trade-off. Cars are always designed with both efficiency and robustness in mind (we are ignoring other design criteria such as comfort, usability, etc., for the sake of simplicity), and one tries to achieve an optimal trade-off between efficiency and robustness for a given cost constraint.

Such a design philosophy is not limited to just automobiles. It is widely used in the design of all human-engineered systems, e.g., airplanes, trains, ships, chemical plants, nuclear plants, electric power grids, airline networks, the Internet, etc. Even nature pays attention to the efficiency-robustness trade-off in its design of biological systems. Efficiency determines the *short-term* survival while robustness dictates the *long-term* survival of a system. The overall survival fitness of a system, whether it is an institution, a society, or a biological organism, is determined by a balance of both efficiency and robustness, as demanded by the environment (Venkatasubramanian et al. 2004; Venkatasubramanian 2007).

Given such a universal design criterion, it is hard to believe that the free-market system, and the capitalist society at large, both of which are dynamical systems routinely subjected to disturbances and shocks, would have evolved, and would operate, by focusing solely on efficiency while ignoring robustness altogether. From a systems engineering perspective, it makes little sense to think that the free-market society would have done this. This author has always been puzzled by this penchant of most mainstream economists who claim that only efficiency matters, while completely ignoring the robustness issue, which is just as vital. Systems engineering tells us emphatically that robustness must matter in free-market functioning. If that is the case, what property (or properties) gives rise to robustness of a free market? Our mathematical analysis shows that it is fairness, which makes intuitive sense. By maximizing fairness, the ideal free market achieves asymptotic stability, thereby achieving robustness, and resilience, to a certain amount of stress.

No system, human-engineered or nature-evolved, is completely robust to all kinds of stress. Any, and every, system can be broken down by applying a sufficient amount of disruption. It is simply prohibitively expensive to design a system that is totally fail-proof. So the best one can do is to achieve an optimal trade-off, as dictated by the operating environment, for the given cost constraint. Furthermore, systems typically have several layers of protection to achieve robustness. Some of the robustness might be due to the inherent nature of the underlying dynamic process, while the rest is achieved by the additional layers such as supervisory control systems and safety interlocks. These kick into action during emergencies due to large upsets or breakdowns. Similarly, despite the inherently stable nature of the free market, we do need additional layers of protection, such as the Securities and Exchange Commission and the Federal Reserve Board, to handle large upsets and shocks.

To summarize, systems engineering principles demand that both efficiency and robustness requirements be addressed in the design and operation of a free-market system. Economists have always known that efficiency mattered. Now we realize robustness also matters. Since this robustness is derived from fairness in distributive justice, we see that fairness must matter in the operation of a free-market system. Thus, we see that our mathematical analysis supports our intuition and historic lessons regarding fairness and stability.

It is interesting to note that Keynes had intuitively recognized the importance of this balance when he remarked (as quoted in the beginning of this book), "The political problem of mankind is to combine three things: economic efficiency, social justice and individual liberty." The social justice component corresponds to robustness in design and operation from a systems engineering perspective. Individual liberty is already accounted for in our theory following the guidelines of Rawls and Nozick. Thus, our theory addresses all three elements highlighted by Keynes.

Now, one could critique that all that we have proved is that the equilibrium income distribution is asymptotically stable. One could protest that to go from this result to making claims about the robustness and the resilience of the entire capitalist society is too much of a stretch.

Perhaps so, but we defend our claim by arguing two reasons. First, the income distribution is at the root of all things good and bad in a capitalist society. This is where "the rubber meets the road." This is the bottom line. When one's income is unfairly low, all kinds of resentment begin to set in. On the other hand, when one is being compensated well in a job, it generally implies that most things are going well, as far as the work environment is concerned; e.g., the worker is being treated fairly in a number of areas such as race, gender, qualification, and contribution.

We do not wish to get into an extensive discussion here on this complex topic as there are many nuances to this line of reasoning. All we wish to stress here is that the income distribution is an extremely important metric of the health of a capitalist society. One's income determines one's wealth, and together they determine a number of other important quality-of-life measures such as education, access to health care and justice, family life, and so on, not only for the immediate household members but also for future generations. Economic ability, i.e., economic freedom, essentially determines the ability of someone to exercise her other freedoms and rights.

Consider, for example, the plight of the bottom 20% of U.S. households, comprising tens of millions of people, who make less than $18,000 per year. Imagine what their lives are like. What kind of quality of life can they enjoy? Most are U.S. citizens, and so legally they enjoy all the rights conferred upon them by the U.S. Constitution. How many of these rights are they able to exercise to enjoy a good quality of life in their pursuit of happiness? For all practical purposes, they lead lives as if they have no rights, no say, in the various spheres of their existence—in their workplace, in their communities, in their governments, and so on. Thus, the income distribution has a direct and significant impact on the three foundational principles of U.S. society—life, liberty, and the pursuit of happiness—for all citizens. It is an excellent proxy for the overall health of a capitalist society. The Dow Jones is not.

Also, the stability of the income distribution is really a verdict on the stability properties (both robustness and resilience) of the underlying dynamic process, namely, the free-market competitive mechanism. Since this is the fundamental basis of a capitalist society, it is reasonable to state that the asymptotic stability proof implies the robustness and resilience of a capitalist society.

One could ask: How relevant is this stability result for real-life free markets which are not maximally fair, as we discussed in sections 1.4 and 3.1? (We are also anticipating our results from chapter 6 a bit here.) Why have they been more or less stable for a couple of centuries in the Western democracies despite being not maximally fair most of the time?

We believe this is due to a combination of two things. First is the *asymptotic stability* property, i.e., the natural tendency to get back to the original, equilibrium, state after taking a hit. We conjecture that this equilibrium can, in fact, be shown to be *exponentially stable*. That is, after a shock, will it not only return to its original equilibrium state but also it will do so very quickly. We think that this stability property is so strong that the society can tolerate a certain level of unfairness and abuse before it begins to fall apart. Second, whenever we reached such levels of unfairness, as noted, structural improvements and reforms have occurred, leading to new systems, laws, and regulations to correct and protect a capitalist society.

As discussed, fairness or robustness generally increases the cost to the system in the near term, as it usually comes at the expense of efficiency. However, in the long run, as fairness *empowers* a vast majority of the population, who were previously excluded from (or restricted in) their fair share of prosperity, their collective contributions to innovation, productivity, and aggregate demand greatly power *economic growth*, thereby increasing everyone's take of a much larger pie in the long run. Thus, increasing fairness is not only the moral thing to do but also the smart thing to do.

Now we take a step back and ask a broader question. How is the stability of this equilibrium related to the Walrasian general equilibrium? We explore this important question next.

## 5.11 Comparison with Walrasian General Equilibrium

It's interesting to contrast the asymptotic stability of this equilibrium with the stability concerns of the general competitive equilibrium in the Walrasian model (Jorgenson 1960; Kirman 1989; Fisher 1989; Ackerman 2002; Kirman et al. 2006; Kirman 2010; Fisher 2011; Kirman 2011; Bassett and Claveau 2011; Helbing and Kirman 2013). It is not our purpose to delve into this deeply here, but it certainly needs to be pursued, for the question of the stability of general equilibrium is a complex and challenging subject in itself. Nevertheless, we believe it is important to raise this comparison to begin to explore and understand its ramifications.

The long-term stability of a democratic, free market–based capitalist society depends on a number of foundational elements. The foremost of these are, as we discussed in chapter 2: (1) equality of basic rights for all citizens; (2) properly functioning institutions to enforce law and order and to provide other essential services (e.g., food, education, health, financial, infrastructure) needed for one's life; (3) a growing economy; (4) the fruits that economy shared among all participants fairly; (5) for all, productive and meaningful lives in the pursuit of happiness. If we assume that elements 1 and 2 are satisfied, then the stability of the society depends on achieving elements 3 and 4, which ought to lead to element 5, at least in economic terms. Our theory proves that in our utopia, BhuVai, element 4 is satisfied at equilibrium, which is also proved to be asymptotically stable and socially optimal.

This leaves element 3 as an open question, which primarily depends on innovation and productivity growth in the economy. Element 3 also depends on the stability of general competitive equilibrium, which appears to be an unsettled question.

In the Walrasian formulation of the general competitive equilibrium, given a number of agents, with certain endowments and demands of different products and with certain product preferences in maximizing their utilities, one is essentially asking, as Fisher (2011, 34) observes: (1) Can such a system have an equilibrium solution? (2) Is it unique? (3) Is it Pareto-optimal? (4) Is it stable?

The first three questions have been answered in the affirmative, under quite general circumstances. Fisher (2011, 34–35) states, "The two welfare theorems (loosely stated above) provide the rigorous justification for the view that free markets are desirable (although they say nothing about fairness, or any other desirable attribute other than Pareto-efficiency). It is not an overstatement to say that they are the underpinning of Western capitalism."

But Fisher (2011, 35) then cautions and emphasizes:

So elegant and powerful are these results, that most economists base their conclusions upon them and work in an equilibrium framework—as they do in partial equilibrium analysis. But the justification for so doing depends on the answer to the fourth question listed above, that of stability, and a favorable answer to that is by no means assured. It is important to understand this point which is generally ignored by economists. No matter how desirable points of competitive general equilibrium may be, that is of no consequence if they cannot be reached fairly quickly or maintained thereafter, or, as might happen when a country decides to adopt free markets, there are bad consequences on the way to equilibrium.

Such concerns have also been articulated by Ackerman (2002, 132–33):

General equilibrium is still dead. Exactly 100 years after the publication of Walras' most important work, the SMD theorem proved that there was no hope of showing that stability is a generic property of market systems. More than a quarter-century of additional research has found no way to sneak around this result, no reason to declare instability an improbable event. These negative findings should challenge the foundations of economic theory. They contradict the common belief that there is a rigorous mathematical basis for the "invisible hand" metaphor; in the original story, the hand did not wobble.

Ackerman (2002, 135) further observes:

Market economies are only episodically unstable or chaotic; it is certainly the norm, not the exception, for markets to clear and for prices to change smoothly and gradually. In short, it is more obvious in practice than in theory that large, complicated market economies are usually stable. If it is so difficult to demonstrate that stability is endogenous to a market economy, perhaps it is exogenous. That is, exogenous factors such as institutional contexts, cultural habits, and political constraints may provide the basis for stability, usually damping the erratic endogenous fluctuations that could otherwise arise in a *laissez-faire* economy.

Thus, we have the intriguing situation in which the theory is unable to prove stability while the system in practice is mostly stable. This is curious because it is often the other way around in engineering systems. The theoretical model might predict stability but is not observed in reality, because the model made certain assumptions about the system that are not valid in practice.

Now, contrast the stability concerns of the Walrasian general equilibrium model with the stability properties we have proved. The Walrasian general equilibrium result essentially states that, at equilibrium, all demands are met, and all supplies are exhausted, for all the products, and the market clears for all of them at their respective prices. There is no excess demand or supply left at equilibrium.

This is essentially what we have proved in our case as well. Here, the different products are the workers with different skills (ranging from a janitor to a CEO), their prices are wages or salaries, and the quantities are the number of workers at each salary level. At equilibrium, in our thought experiment, all $N$ workers found their salary levels, with no worker left behind and no skill level (i.e., product category or salary level) unfilled; i.e., the market cleared for everyone and every category.

Therefore, the equilibrium income distribution can be thought of as the general competitive equilibrium outcome for the labor market. In addition, we have proved that this equilibrium is asymptotically stable; i.e., after a disruption, the distribution will actually return to its original equilibrium state.

As we know, the overall structure of the income distribution has been largely stable for over a century in different countries. It is lognormal for the bottom 95% to 97% of the population and a Pareto distribution for the top 3% to 5%. Our theory predicts lognormal for the entire population for a one-class society and two different lognormals for a two-class society (we discuss this next, in section 5.12), where the top lognormal could be easily misidentified as a Pareto distribution (we discuss this in section 6.7). Thus, we have shown, through analytical work and agent-based simulation (see section 6.7), that the self-organization process (a bit like the *tatonnement*) can start far away from equilibrium (e.g., from a delta function of salaries, initially, in our thought experiment) and get to the equilibrium distribution quickly, thus verifying the stability of the equilibrium to disruptions.

It is therefore quite puzzling that we are able to prove the uniqueness, optimality, and asymptotic stability of the equilibrium income distribution without too much trouble in an ideal free market (and without needing to invoke any *exogenous factors* as Ackerman recommends above), invoking only the self-organizing invisible hand process (which is similar to the Walrasian *tatonnement* process), while under the same ideal free-market conditions, the stability of the general equilibrium remains a questionable outcome. This is particularly perplexing because one could think of our result as proving general competitive equilibrium and its stability in the context of labor markets.

We need to understand this apparent conflict more deeply, intuitively, and analytically. What is the relationship between these two equilibria, the Walrasian and the income distribution? How does one affect the other? And how do they affect the overall stability of a free market–based capitalist society in practice?

One of the main causes of this apparent conflict, we believe, is the Newtonian classical mechanics–like framework of the Walrasian general equilibrium model. As we explained in chapter 4, such an approach was not successful in the case of thermodynamics, and it required an entirely different formulation, namely, statistical mechanics. Similarly, we believe the stability properties of the economy should be addressed by using a game theoretic–statistical mechanics approach similar to the one we proposed. This has served us well by addressing this problem in the context of labor markets—by proving the existence, uniqueness, optimality, and stability of the equilibrium, all the wonderful properties we so desire and hope for. The next step would be to extend this approach to the general equilibrium of all commodities, not just labor. While this author has made some progress regarding this, we don't discuss it here as the work is presently incomplete.

On the other hand, some have questioned whether the economy operates at or near equilibrium in practice at all (Minsky and Kaufman 1986; Keen 2011). This line of inquiry has gained particular attention since the 2007–2009 Global Financial Crisis—what some economists call a Minsky moment (Cassidy 2008).

At any rate, we need more analytical, simulation, and empirical studies to answer these important questions, particularly from the perspective of statistical mechanics and dynamical systems engineering. These efforts are still in infancy, often advocated by econophysicists, and are viewed with a great deal of suspicion, if not outright derision, by mainstream economists.

Alan Kirman (2011) and Philip Mirowski (1989) offer valuable insights regarding why mainstream economics veered away from its origins in physics toward axiomatic, Bourbaki-style mathematics. Kirman (2011, 127) points out: "I would now like to offer a final explanation as to why we have pursued the road from Walras to Arrow-Debreu so assiduously, and it concerns the relationship between economic theory and physics and mathematics. Mirowski (1989) has built a painstaking and convincing argument that Walras's theory was built directly on the basis of classical mechanics." Later on, in the same section, Kirman (2011, 128) observes: "If, as seems clear, the Lausanne School was built on physics why did it not evolve with that science rather than evolve towards pure mathematics? Mirowski's (1989) explanation is clear. Had it done so it would have found itself in a contradictory position. Thermodynamics and the emphasis on entropy would suggest a system which was constantly moving towards disorder, the opposite of the Lausanne view." We see, again, how entropy has been a major stumbling block for over a century at the very foundations of economics. We believe we have removed this conceptual hurdle with our reinterpretation of entropy as a measure of fairness.

Commenting on economics pivoting toward abstract mathematics, Kirman (2011, 129) suggests: "In my view, in the 1950s, theoretical economists, led by Debreu, turned their backs on physics and tried to build an abstract coherent structure in which equilibrium states could be shown to exist. Economics was, at least in the view of those in mathematical economics at the time, ready to build a structure as minimalist and abstract as that provided by Bourbaki for mathematics. . . . Physics envy was thoroughly replaced by mathematics envy."

Kirman (2011, 130) concludes: "The desire to be mathematically rigorous is still strong, but the abandonment of the physics approach in favour of the axiomatic approach has made at least one alternative route, that of statistical physics less accessible and, what is more, unwelcome."

I believe that economics, at its core, is an applied field like engineering dealing with practical matters of everyday life of people in a society. It ultimately needs to deal with the realities of systems and phenomena in the world of commerce for it to be relevant, much as engineering has to in its various real world applications. Mathematics plays a crucial role in all this by trying to bridge theory and reality, by sharpening and guiding one's initial intuition and reasoning through appropriate degrees of abstraction, formalization and model formulation, to explain and predict real world phenomena. In this sense, we believe, mathematical tools are to be used in economics in a manner similar to their applications in engineering and applied physics, as opposed to the Bourbaki-style formal mathematics that has come to dominate mathematical economics over the past several decades. As a result, as Ackerman, Fisher, Foley, Kirman, Minsky, Mirowski, and others have argued for a long time, there are no satisfactory answers to certain fundamental questions in mainstream economics. So there is no other option but to look elsewhere for different conceptual frameworks. This is what we have attempted in this book.

## 5.12  Bipopulation Game

Our model with a homogeneous population of agents with the same payoff preferences (i.e., same $\alpha$, $\beta$, and $\gamma$ values for every agent) describes a one-class system, i.e., a classless society—utopia. At the other extreme, different agents would have different payoff preferences, i.e., different $\alpha$, $\beta$, and $\gamma$ values. However, reality is typically somewhere in between, with $\Pi$ different classes of agents, with all the agents in the same class having the same $\alpha$, $\beta$, and $\gamma$ values. If they do not have the same values, they have values spread around mean $\alpha$, $\beta$, and $\gamma$ values for each class. For instance, employees in an organization, or in a society at large, are often grouped into three broad classes: [blue collar, white collar, C-suite executives or owners] or [lower, middle, and upper class]. Such coarse-grained classification is often appropriate, even necessary sometimes, to elicit and discern macroscopic trends in a population. With that in mind, we now present the analysis for a two-class (i.e., bipopulation) system containing two classes of agents, each with a distinct set of $\alpha$, $\beta$, and $\gamma$ values. We then show how this can be generalized to $\Pi$-class systems.

The effective utility of an agent in a two-class system at salary level $i$ is therefore defined as

$$h_{i,j} = \alpha_j \ln S_i - \beta_j (\ln S_i)^2 - \gamma_j \ln(N_{i,1} + N_{i,2}) \qquad (5.42)$$

where the choice of $j \in \{1, 2\}$ indicates either class 1 population or class 2 population.

The equilibrium replicator dynamics is also modified:

$$\begin{aligned} h^*_{i,j} &= h^*_j \quad \forall i \in \Omega_j \\ h_{k,j} &< h^*_j \quad \forall k \notin \Omega_j \end{aligned} \qquad (5.43)$$

where $\Omega_j = \{k | x^*_{k,j} > 0\}$ denotes the collection of levels with class $j$'s presence. The first condition is identical to the homogeneous scenario. The second indicates the possibility that some levels are occupied by only a single class (i.e., the utility is too low for the other class).

We can prove that $\Omega_1 \cup \Omega_2 = \{k | 1 \leq k \leq n\}$ and $\Omega_1 \cap \Omega_2 = \emptyset$; i.e., every salary level contains some population but not both. First, suppose there are empty salary levels. They will soon be occupied because of the infinitely high utilities. Thus $\Omega_1$ and $\Omega_2$ cover the whole domain. Second, suppose there is an overlap where $\hat{\Omega} = \Omega_1 \cap \Omega_2$. Let the equilibrium population density be $x^*$. Equation (5.43) can be rewritten as

$$h^*_j = \alpha_j \ln S_i - \beta_j (\ln S_i)^2 - \gamma_j \ln N x^* \qquad \forall i \in \hat{\Omega} \qquad (5.44)$$

Thus

$$\lim_{\Delta S \to 0} \frac{\Delta x^*}{\Delta S_i} = \frac{dx^*}{dS} = \frac{\alpha_j - 2\beta_j \ln S}{\gamma_j S} x^* \qquad (5.45)$$

This indicates two distinct gradients for *every* point in $\hat{\Omega}$. Therefore we will not see an overlapping region with mixed populations.

We can also prove that the equilibrium density curve is continuous at the interface of two populations. Suppose otherwise, at the interface $S = \hat{S}$,

$$x_1^*(\hat{S}) \neq x_2^*(\hat{S}) \tag{5.46}$$

According to equation (5.43) again,

$$\alpha_j \ln \hat{S} - \beta_j (\ln \hat{S})^2 - \gamma_j \ln x_j^* \geq \alpha_j \ln \hat{S} - \beta_j (\ln \hat{S})^2 - \gamma_j \ln x_{-j}^* \tag{5.47}$$

that is,

$$x_j^* \geq x_{-j}^* \tag{5.48}$$

The only possible solution is $x_1^* = x_2^*$; therefore, the population density is continuous at the interface.

Even though we now know these equilibrium characteristics of a bipopulation game, an exact equilibrium density is still tedious to obtain unless $\Omega_1$ and $\Omega_2$ are given:

$$\begin{aligned} x_i = & \frac{N_1/N}{S_i D_1} \exp\left[-\frac{\left(\ln S_i - \frac{\alpha_1 + \gamma_1}{2\beta_1}\right)^2}{\gamma_1/\beta_1}\right] \mathbb{1}(i \in \Omega_1) \\ & + \frac{N_2/N}{S_i D_2} \exp\left[-\frac{\left(\ln S_i - \frac{\alpha_2 + \gamma_2}{2\beta_2}\right)^2}{\gamma_2/\beta_2}\right] \mathbb{1}(i \in \Omega_2) \end{aligned} \tag{5.49}$$

where $D_j$ is the normalization that ensures $\sum_{i=1}^{n} x_i = 1$. We can, however, get a good approximation when the two lognormal curves of class 1 and class 2 are sufficiently separated that the overlap is insignificant. The overall distribution is then estimated as a mixture of two lognormal distributions:

$$x_i \approx \frac{N_1}{S_i D} \exp\left[-\frac{\left(\ln S_i - \frac{\alpha_1 + \gamma_1}{2\beta_1}\right)^2}{\gamma_1/\beta_1}\right] + \frac{N_2}{S_i D} \exp\left[-\frac{\left(\ln S_i - \frac{\alpha_2 + \gamma_2}{2\beta_2}\right)^2}{\gamma_2/\beta_2}\right] \tag{5.50}$$

where $N_j$ denotes the number of class $j$ agents and $D$ denotes the normalization parameter, which is easily computed.

## 5.13  Midpoint Summary

As we near the end of this chapter, we are a bit past the halfway point of the book, which is a good place to review the progress so far. Piketty, Atkinson, Saez, and others have performed yeoman's service by carefully documenting empirical data that clearly demonstrate the increasing inequality in income and wealth in several countries in recent years. As we discussed, people realize *intuitively* that the current level of inequality is unfair, is morally and economically unjustified, and is potentially destabilizing. This extreme inequality negatively impacts a number of critical economic and social factors, such as growth and efficiency, access to health care, to education, and to justice, as well as opportunities for upward mobility.

While the qualitative arguments made on both sides of the inequality debate are interesting, illuminating, and useful to highlight the importance of this issue, they nevertheless lack the analytical and quantitative insights one could gain, and one needs, from a mathematical theory of fairness in a capitalist society.[8]

In particular, from a theoretical perspective, we would like the following fundamental questions answered quantitatively: (1) What kind of income distribution should arise naturally, under ideal conditions, in a free-market environment comprising utility-maximizing employees and profit-maximizing companies? (2) Is this distribution fair? (3) Is a free-market society morally justified, at least under ideal conditions? (4) Is a free-market society stable, at least under ideal conditions?

As it turns out, there are no answers to these fundamental questions in conventional economic theories. While there have been some attempts to address the first question in econophysics models, these approaches are counterintuitive to well-established economic principles. Furthermore, and most crucially, they do not address the fairness, morality, optimality, and stability questions.

In this book, from chapter 2 through chapter 5, we developed an unorthodox theoretical framework that integrates key concepts from political philosophy, economics, game theory, statistical thermodynamics, information theory, and systems engineering to answer all these questions mathematically.

In particular, we proved that the fairest income distribution is lognormal, achieved at equilibrium, in an ideal free-market society. We dispelled the common misinterpretations of entropy as randomness, disorder, or uncertainty, and we showed that it actually is a measure of fairness in a distribution. Building on the deep insight that game-theoretic potential and entropy are the same, we proved that Nash equilibrium is arrived at by maximizing fairness in a distribution. We also proved that at this equilibrium all workers enjoy the same level of effective utility (i.e., economic welfare or happiness) and that this is also the socially optimal outcome (i.e., this is the best a society can deliver to its working citizens). Both these results provide the moral justification for an ideal free-market economy, mathematically proved for the first time. This is the reason for everyone to voluntarily participate in a capitalist society. If you participate and contribute in a free-market economy, under ideal conditions, you will not be left behind. This is the moral basis for the underlying social contract and social stability.[9] We also show how this result fits well with the concept of robustness in systems engineering, which is balanced against efficiency in system-theoretic approaches to the design of engineering systems.

We also contrasted our asymptotic stability proof with the inability of the Walrasian general equilibrium model to achieve a similar outcome. This difference could possibly be due to the reliance of the Walrasian model on the concept of mechanical equilibrium, by invoking the notions of supply and demand forces from classical mechanics, whereas in our approach we utilize the concepts and tools of statistical mechanics by recognizing that the free-market dynamics leads to a statistical equilibrium—i.e., this outcome is not a mechanical equilibrium, but more like a chemical equilibrium.

This doesn't mean one should use our framework for all problems in economics. Even though macroscopic matter is made up of atoms, we don't utilize the atomic perspective, that level of granularity, for all problems in physics and engineering. When we design bridges the atomic perspective is not useful and hence not appropriate. There, one uses a macroscopic model of matter to calculate the forces, etc. Similarly, when we model liquid flow through pipes we don't use the atomic view. Navier-Stokes equations for modeling liquid flow do not invoke atomic theory of liquids. However, in many situations in thermodynamics, e.g. modeling phase transitions and critical phenomena, the atomic perspective, i.e., the statistical mechanics perspective, is not only appropriate but indispensable. But even in thermodynamics, we don't always use this perspective. There are many engineering applications involving mass and heat transfer, as in the design of steam engines, reactors, and distillation columns, where that level of detailed modeling is unnecessary and unproductive. Thus, the nature of the problem dictates which level of abstraction and modeling are appropriate. Similarly, one doesn't need to invoke the statistical teleodynamics framework for all problems in economics. For certain problems, the neoclassical economic perspective is both appropriate and effective.

Our utopian free-market society is founded on the Principle of Maximum Fairness, which really is a statement about the fundamental equality of all individuals. This is the real message in the Golden Rule: "Do unto others as you would have them do unto you." One might think of the set of equations 3.6, 3.11–3.15, 5.21, 5.25, 5.27, and 5.34, as the mathematical version of the Golden Rule in the context of distributive justice. This author is a firm believer in Lord Kelvin's maxim cited at the beginning of chapter 4: "Do not imagine that mathematics is harsh and crabbed, and repulsive to common sense. It is merely the etherealisation of common sense." The aforementioned set of equations show how the eminently agreeable common sense of the Golden Rule can be expressed mathematically to gain practical guidance on a central problem in political economy. Such aphorisms as the Golden Rule are not just cliches to be dismissed as tired old sayings of little practical value. On the contrary, they are pregnant with great insights, which can be of immense practical utility in addressing challenging problems in economics and sociology, if only one knows how to formulate them mathematically. However, in this pursuit, one should avoid, as noted, the Bourbaki-style mathematical approaches that disregard important practical considerations in favor of mathematical rigor. To paraphrase Einstein, mathematics without conceptual insights is blind, conceptual insights without mathematics is lame.

Recognizing and emphasizing this dormant power of *qualitative* insights, the inimitable Richard Feynman, one of the legendary theoretical physicists of the twentieth century, remarked:

> If, in some cataclysm, all of scientific knowledge were to be destroyed, and only one sentence passed on to the next generation of creatures, what statement would contain the most information in the fewest words? I believe it is the atomic hypothesis that all things are made of atoms—little particles that move around in perpetual motion, attracting each other when they are a little distance apart, but

repelling upon being squeezed into one another. In that one sentence, you will see, there is an enormous amount of information about the world, if just a little imagination and thinking are applied. (Feynman 2011, 4)

Indeed, our theory is such an attempt to find the appropriate mathematical formalism that exploits the central insights of John Locke, Adam Smith, Jeremy Bentham, John Stuart Mill, John Rawls, Robert Nozick, and Ronald Dworkin to address the challenge of distributive justice, *quantitatively* and *analytically*.

Now that we have mathematically identified the gold standard of fairness, the lognormal distribution, we can compare this ideal outcome with real-life income distributions from various countries as well as with data from computer simulations. We can now assess which countries are treating its workers fairly and which are not. We can quantify how fair the income inequality is in a given country. This information could be used to develop better tax and transfer policies to correct the inequities. This is what we will do in the next chapter.

CHAPTER SIX

# Global Trends in Income Inequality: Theory Versus Reality

> All for ourselves and nothing for other people, seems, in every age of the world, to have been the vile maxim of the masters of mankind. As soon, therefore, as they could find a method of consuming the whole value of their rents themselves, they had no disposition to share them with any other persons.
> ADAM SMITH

> The man of great wealth owes a peculiar obligation to the state because he derives special advantages from the mere existence of government.
> THEODORE ROOSEVELT

As discussed in chapter 5, the lognormal distribution is the fairest income distribution that guarantees that all citizens in BhuVai are equally happy, at equilibrium, as they lead their lives the way they wish, while making a valuable contribution to society. No one in this utopia would want to trade his life for someone else's, as he is already as happy as everyone else. A society where everyone is equally happy is the best that a society can deliver to its citizens. What more could one ask of a society? It has achieved its purpose. This is the utopia everyone would love to live in. This is the target society we should all strive for. We can't do any better than this. So if a real-world society comes close to this utopia, it is doing very well for its citizens.

How close are the real-world societies to BhuVai? That is, how close are the real-world income distributions to the ideal lognormal one?

To answer this, we compare the prediction made by our theory with *pretax* income data, as reported by Piketty and his coworkers, for a dozen different countries. We then define a new measure of income inequality, $\psi$, which quantifies how fair a given country is with respect to sharing the total income pie among its citizens. Now $\psi$ reveals which countries are closer to BhuVai and which are not. Since $\psi$ identifies the target (i.e., fairest inequality) *quantitatively*, we could use it to rationally formulate and fine-tune tax and transfer policies that result in a more equitable society. We also compare our predictions for a two-class society with the results from agent-based simulations.

## 6.1 Comparing Theory with Reality

While our theory is developed for modeling pay distributions in large corporations, its predictions may be compared with countrywide pretax income (excluding capital gains) data as most of it is an aggregate of the pay of individuals. In essence, we are approximating an entire country as a large corporation functioning in a free-market environment. We compare our model's predictions for the shares of the total income by three segments of the population (namely, bottom 90%, top 10% to 1%, and top 1%) with those observed in different countries as reported by Piketty and his colleagues in their World Top Incomes (WTI) Database (Alvaredo et al. 2015). Ideally, we should be comparing our predictions with wage income data. However, since we weren't too confident about the reliability of what we found for a number of countries, we decided to use the pre-tax income data, excluding capital gains, from the WTI Database. Since this data includes dividend income (according to the information we received), the inequality effect might be somewhat more pronounced. However, since we are using the same basis across different time periods, and across different countries, our main conclusions regarding the key trends and outcomes remain the same.

While there are different statistical procedures and measures to use to test the closeness of any two distributions, we focus on the one that matters most—the amount of money in the income pie shared by different classes. We compare the income shares of the top 1%, top 10% to 1%, and the bottom 90% of the population in both cases, BhuVai versus the real world. This is, after all, what the income inequality debate is all about. Of course, we fully expect real-life free-market societies to deviate from ideality, but we are interested in understanding how big the deviations are and why.

While the income distributions should be lognormal in different countries, if they were functioning as BhuVai, they may have different means and variances depending on the minimum, average, and maximum annual incomes in the different countries, which determine the $\mu$ and $\sigma$ of the corresponding lognormal distributions. As an example, we show these parameters in table 6.1 for the different countries.

The minimum, average, and maximum income data are obtained from the WTI Database. For the maximum income, we chose to use the threshold income at 99.9%. At this cutoff, the area under the lognormal curve would correspond to $6\sigma$. Strictly speaking, $6\sigma$ covers 99.73% of the population in general, but in our case it covers 99.865% as there is no one below the minimum-income threshold. So this is a very good approximation. Thus, we estimate $\sigma$ by using the approximation

$$6\sigma \approx \ln(\text{maximum income}) - \ln(\text{minimum income}) \qquad (6.1)$$

$$\mu = \ln(\text{average income}) - \frac{\sigma^2}{2} \qquad (6.2)$$

For the United Kingdom and Netherlands where the 99.9% threshold data are not available, we found that the data are well approximated by the average salary of the top 0.5%, by testing this heuristic for the other countries where the threshold is known. For Switzerland, Sweden, Norway, and Denmark, there is no minimum-wage requirement and hence those data are not available. For these countries, we consulted several country-specific sources to obtain guidelines about what the typical minimum wage–like compensation might be for entry-level positions in recent years (2010 to 2014). We then used historic data on annual increases for the average income of the bottom 90% of the population to deflate and back-calculate the minimum wage for the past years.

Once $\mu$ and $\sigma$ are known, we can uniquely determine the corresponding lognormal distribution and compute the income shares of the bottom 90%, top 10% to 1%, and the top 1%. Since $\mu$ and $\sigma$ typically vary from year to year (because of the changes[1] in the minimum, average, and/or maximum income from year to year), the ideal distribution of income to the bottom 90%, top 10% to 1%, and the top 1%,

TABLE 6.1
Model Predictions and Sample Empirical Data for Different Countries

| Country | Norway | Sweden | Denmark | Switzerland | Netherlands | Australia | France | Germany | Japan | Canada | United Kingdom | United States |
|---|---|---|---|---|---|---|---|---|---|---|---|---|
| Year | 2011 | 2012 | 2008 | 2010 | 1999 | 2010 | 2006 | 2008 | 2010 | 2000 | 2011 | 2013 |
| Currency | NOK | SEK | DKK | CHF | NLG | AUD | EUR | EUR | KJPY | CAD | GBP | USD |
| Minimum | 139,152 | 159,630 | 146,919 | 37,266 | 28,142 | 29,716 | 15,885 | 17,927 | 1,236 | 11,036 | 11,324 | 15,131 |
| Average | 319,391 | 249,760 | 202,502 | 67,056 | 63,557 | 49,304 | 26,872 | 29,826 | 2,255 | 24,859 | 19,217 | 52,619 |
| Maximum | 3,874,714 | 2,580,353 | 1,747,643 | 1,047,750 | 416,931 | 688,700 | 366,769 | 630,374 | 32,608 | 528,156 | 365,130 | 1,424,300 |
| $\mu$ | 12.52 | 12.32 | 12.13 | 10.96 | 10.96 | 10.67 | 10.06 | 10.13 | 7.57 | 9.91 | 9.70 | 10.58 |
| $\sigma$ | 0.55 | 0.46 | 0.41 | 0.56 | 0.45 | 0.52 | 0.52 | 0.59 | 0.55 | 0.64 | 0.58 | 0.76 |
| Gini coefficient | 0.30 | 0.26 | 0.23 | 0.31 | 0.25 | 0.29 | 0.29 | 0.33 | 0.30 | 0.35 | 0.32 | 0.41 |
| Bottom 90% ideal | 76.6% | 79.3% | 80.8% | 76.6% | 79.7% | 77.6% | 77.6% | 75.4% | 76.9% | 73.8% | 75.9% | 70.0% |
| Top 10% to 1% ideal | 19.5% | 17.5% | 16.5% | 19.6% | 17.2% | 18.9% | 18.8% | 20.4% | 19.3% | 21.6% | 20.1% | 24.2% |
| Top 1% ideal | 3.8% | 3.1% | 2.8% | 3.8% | 3.0% | 3.6% | 3.6% | 4.2% | 3.7% | 4.6% | 4.0% | 5.8% |
| Bottom 90% data | 71.7% | 72.1% | 73.8% | 66.5% | 71.9% | 69.0% | 67.2% | 60.5% | 59.5% | 58.9% | 60.9% | 53.0% |
| Top 10% to 1% data | 20.5% | 20.8% | 20.1% | 22.9% | 22.7% | 21.8% | 23.9% | 25.6% | 31.0% | 28.0% | 26.2% | 29.5% |
| Top 1% data | 7.8% | 7.1% | 6.1% | 10.6% | 5.4% | 9.2% | 8.9% | 13.9% | 9.5% | 13.2% | 12.9% | 17.5% |
| Bottom 90% $\psi$ | −6.5% | −9.1% | −8.6% | −13.2% | −9.8% | −11.0% | −13.4% | −19.8% | −22.6% | −20.2% | −19.8% | −24.3% |
| Top 10% to 1% $\psi$ | 5.1% | 18.4% | 22.2% | 16.7% | 31.7% | 15.7% | 26.7% | 25.6% | 60.3% | 29.7% | 30.5% | 21.9% |
| Top 1% $\psi$ | 104.2% | 128.1% | 117.4% | 177.3% | 77.8% | 156.6% | 150.5% | 234.3% | 153.8% | 184.3% | 221.0% | 200.7% |

as predicted by the model, also shifts from year to year, though not by much. As an example, these values are displayed in table 6.1 for the twelve countries, which are commonly used examples (Piketty and Saez 2003), for the years shown.

## 6.2  A New Measure of Fairness in Income Inequality: $\psi$

Now we know from empirical data that the free-market economies of the Scandinavian countries are generally fairer in the economic treatment of all their citizens, not just the wealthy ones. We also know that the United States does not do as well. We further know that other Western European countries such as France, Germany, and Switzerland are somewhere in between.

Can our theory predict these outcomes? That is, just by knowing only the minimum, average, and maximum incomes in a free-market society, can the model identify how fair these societies are?

To test the model along these lines, we define a new index of inequality that uses the ideal lognormal distribution (one that is appropriate for the country under consideration) as the reference. This new measure, which we call *nonideal inequality coefficient* $\psi$, is defined as

$$\psi = \frac{\text{actual share}}{\text{ideal share}} - 100\% \qquad (6.3)$$

where $\psi$ measures the level of nonideal inequality in the system. When $\psi$ is zero, the system has the ideal level of inequality, the fairest income inequality found in BhuVai. When $\psi$ is small, the level of inequality is almost ideally fair; when it is large, the inequality is more extreme and unfair. We computed the predicted income shares for the different countries for the period from ~1920 to ~2012, depending on the availability of the data in the WTI Database.

We then computed $\psi$ for the three segments (bottom 90%, top 10% to 1%, and top 1%) for these countries, annually, for the corresponding time periods (sample values are shown in table 6.1). These are plotted in figures 6.1 through 6.12. If a country were functioning as an ideal free-market system, as defined by our theory, the corresponding $\psi$ values for the three segments would all be exactly zero. This is the reference line which is shown as the 0% line (black dotted line) in the plots.

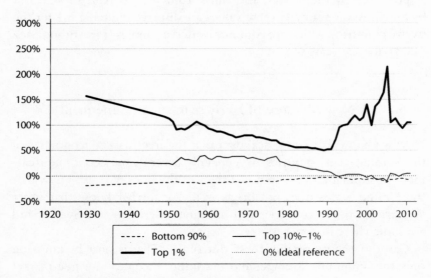

*Figure 6.1* Income inequality in Norway, 1930 to 2010.

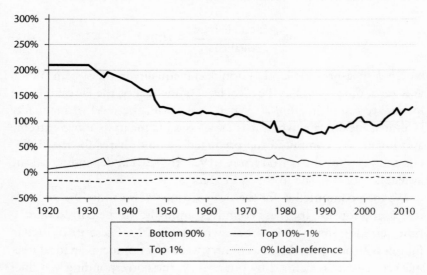

*Figure 6.2* Income inequality in Sweden, 1920 to 2010.

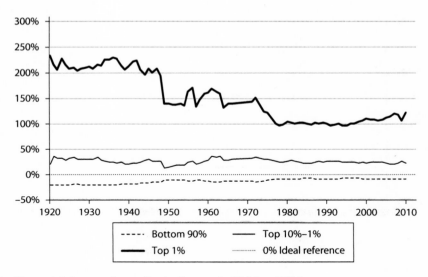

*Figure 6.3* Income inequality in Denmark, 1920 to 2010.

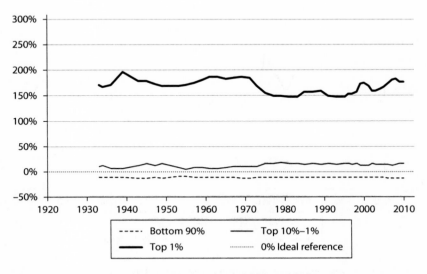

*Figure 6.4* Income inequality in Switzerland, 1930 to 2010.

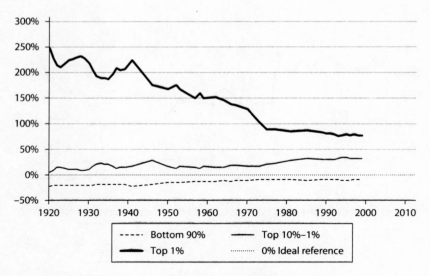

*Figure 6.5* Income inequality in the Netherlands, 1920 to 2000.

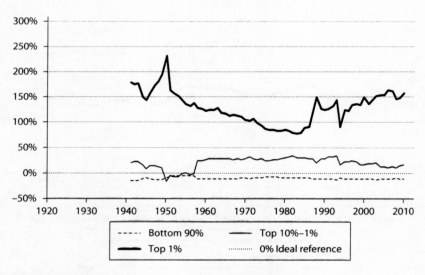

*Figure 6.6* Income inequality in Australia, 1940 to 2010.

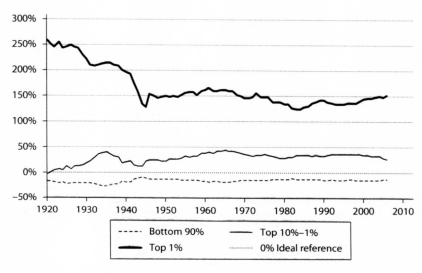

*Figure 6.7* Income inequality in France, 1920 to 2000s.

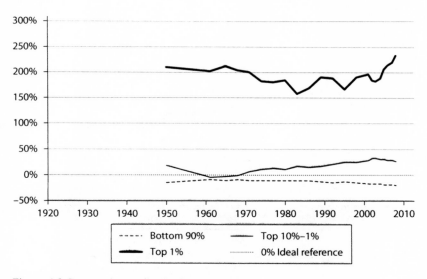

*Figure 6.8* Income inequality in Germany, 1950 to 2010.

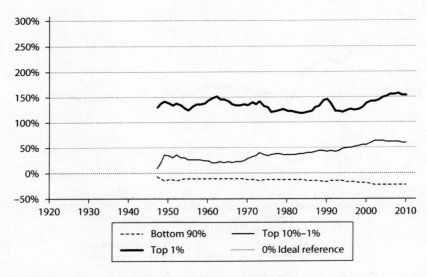

*Figure 6.9* Income inequality in Japan, 1940 to 2010.

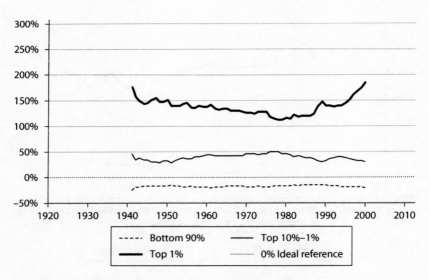

*Figure 6.10* Income inequality in Canada, 1940 to 2000.

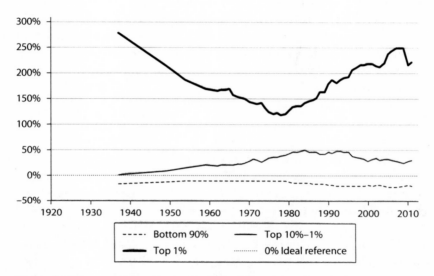

*Figure 6.11* Income inequality in the United Kingdom, 1930 to 2010.

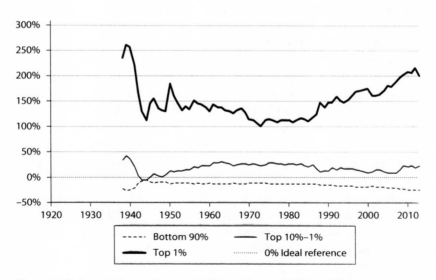

*Figure 6.12* Income inequality in the United States, 1930 to 2010s.

As we can see, the theory's predictions are in general agreement with what is known about these countries regarding their inequalities. The twelve countries are shown, roughly, in the order of generally increasing inequality according to our model. Our objective here is not to rank them in a strict order but to show how different countries have deviated from ideality.

## 6.3 Scandinavia: Almost Utopian

Let's examine what these charts tell us. As we expected, none are ideal, but there are some pleasant surprises. Consider Norway as an example. Its bottom 90% and top 10% to 1% income shares have been remarkably close to the ideal BhuVai values over the last ~20 years.

Its bottom 90% $\psi$ has steadily improved, from a low of about −20% in 1929 to about −2% in 1993, and it has been ~5% to 10% below the ideal value in the last ~20 years. Similarly, its top 10% to 1% $\psi$ has come down from a high of ~40% in 1968 to ~5% in 2011. In fact, during ~1991 to 2011, it has been hugging the ideal line quite closely, sometimes a little bit above and sometimes a little below, typically within a narrow ±6% band.

As for the top 1%, $\psi$ has steadily improved toward the ideal value over a period from ~1929 (at ~160%) to ~52% in ~1990. After spiking up in 2005, it came down to ~104% in 2011. But clearly the top 1%'s share is much more nonideal than that of the bottom 99%.

We find similar close-to-ideality trends in Sweden, Denmark, and Switzerland, for the bottom 90% and top 10% to 1%. In these countries, typically, the bottom 90% $\psi$ is within ~10% of the ideal value; the top 10-1% $\psi$ is within ~15% to 25%.

All these countries, which practice hybrid free-market economies, are generally known to be fairer in their economic treatment of all their citizens. But how fair are they? We could not answer this question before, as there was no reference to compare with, but now we can as our theory provides such a benchmark—the lognormal distribution of BhuVai. We find it remarkable that the income sharing values these countries have achieved for the bottom 90% and top 10% to 1% are so close to the ideal distributions.

We find this to be a surprising result because we didn't expect any real-world economic system to come this close to BhuVai, given the simplifying assumptions and approximations in our model regarding the hybrid utopia. For an overwhelming majority of the population (~99%), these countries have achieved a near-ideal degree of fairness, presumably through an enlightened combination of individual, corporate, and societal values, and macroeconomic policies, all executed through the hybrid free-market mechanism.

Clearly, this did not happen quickly, and it took some time, as the trends show, but it is encouraging to find that through the mistakes made and lessons learned, societies can evolve and adapt to "discover" a near-ideal distribution, given a chance through the political process. In addition, what is even more remarkable is that these hybrid free-market economies did not know, a priori, what the ideal, theoretically fairest, distribution was, and yet they seem to have "discovered" and maintained a near-ideal outcome empirically on their own. While these agreements with the aggregate model predictions are encouraging, more thorough studies are needed, using detailed distribution data to validate these initial impressions and understand the comparisons more rigorously.

From the charts, it appears that the Netherlands, Australia, and France are broadly in the same general class of higher inequality compared to the first group. The next group comprises Germany, Japan, and Canada, and finally the United Kingdom and United States are about the same. It is curious, though, that Japan shows a much higher share for the 10% to 1%, compared with even the United States or United Kingdom. It will be interesting to understand why and how this happens in Japan. Again, our objective here is not to rank these countries in any strict order, but to show how different countries deviate from ideality.

### 6.4 Potential Loss of Growth due to Extreme Inequality

Another interesting result is that from ~1945 to 1975 the United States was only ~12% below the ideal level for the bottom 90%. But, since then, it has lost a lot of ground ending at ~24% below the ideal level in 2012. It is also interesting that the top 10% to 1% dropped from a high of ~30% in 1963 to a low of 8% in 2007. While these two segments lost ground (strictly speaking, the top 10% to 1% is still doing well, enjoying more than its fair share of income), the top 1% went from a low of ~100% in 1973 to a high of ~215% in 2012.

This outcome is what Reich describes as *predistribution* (Reich 2015, xiv): "This has resulted in ever-larger upward predistribution *inside* the market, from the middle class and poor to a minority at the top. Because these predistributions occur inside the market, they have largely escaped notice."

This predistribution, by the way, is not a trivial amount. Our theory helps us estimate the income share lost by the bottom 90%. For instance, in the United States, the bottom 90 percent's fair share is ~70%, but its actual share is ~53%. We realize that real systems cannot approach the ideal value, but if we got back to where we were during 1945 to 1975, at ~63%, it would add a considerable amount of income, an extra 19% (from 53% to 63%), to the bottom 90%.

Doing a quick back-of-the-envelope estimation, in 2010, we found the total wage income in the United States was ~$8 trillion. The fair share for the bottom 90% is ~$5.6 trillion. Instead it received ~$4.2 trillion, a loss of ~$1.4 trillion. If its share had been ~63%, it would have received ~$5.0 trillion, or, $0.8 trillion more—as though every American in the bottom 90% had gotten a 19% raise that year. Given that the top 1% saves ~40%, the next 10% to 1% saves ~12%, and the bottom 90% has near-zero savings, it is reasonable to estimate that ~30% of $0.8 trillion would have been the *extra consumption* (due to the increase in aggregate demand from the expenditures of the bottom 90% over what would have been spent by the top 1% and the 10% to 1% segments) added to the economy. The United States GDP in 2010 was ~$14 trillion. Assuming a multiplier of 1.5, the extra consumption could potentially have increased the GDP by ~$0.36 trillion (or $0.3 \times 0.8 \times 1.5 = 0.36$), *about a 2.5% increase every year.*

In an era when the economy is struggling to grow at 2% per year, an additional growth of about 2.5% is not a trivial amount. Thus, every year the United States has been potentially losing a growth of perhaps ~2% for a decade or more because of reduced aggregate demand from the bottom 90%. Hence, as many economists have argued, besides the troubling unfairness issue, there are the lost opportunities from growth loss that we need to be concerned about. Conservatives tend to emphasize the need for higher economic growth, but have often missed the importance of more widely distributed growth. These estimates stress the potential negative impact of unfair income inequality on economic growth. Admittedly, these are very rough estimates, and one should perform more careful calculations, but it's likely that we are losing at least 1-2% growth annually due to extreme income inequality.

Economists have known that the period from about 1945 to 1975 was when both the bottom and middle classes were doing well in the United States, but we see here how well they were doing in sharing the fruits of the country's progress. While the country might have been more unjust in racial and gender inequalities in that period than now, it appears to have been closer to ideality in economic matters. Our theory, in particular $\psi$, could provide quantitative guidance for formulating tax and transfer policies, which can correct for these inequities in after-tax income, such that after-tax $\psi$ is close to zero, thus restoring fairness in the society.

## 6.5 Are the Rich Different from You and Me?

F. Scott Fitzgerald, the great American novelist who chronicled the lifestyle of the rich and famous in the Roaring Twenties in his magnum opus, *The Great Gatsby*, is reputed to have said, "The rich are different from you and me." In reply, Ernest Hemingway, a literary giant himself, who won the Nobel Prize in Literature in 1954, is supposed to have quipped, "Yes, they have more money."

Indeed! And our analysis shows how much more. We see that in all these countries, including Scandinavia, the top 1% have deviated from ideality the most. They are enjoying much more than their fair share according to our theory.

What if they are indeed different and better? If we accept that in real life the top 1% are indeed different from the rest of the population (i.e., we have a two-class society) because of their special talents and drive, then one could argue that they should get more as they contribute more.

But how much more?

We find that perhaps Scandinavia can help us answer this question. The Scandinavian countries, which have managed to approach a near-ideal distribution empirically for the bottom ~99%, seem to allocate about ~50% to 100% more than the ideal share for the top 1% (see figures 6.1 through 6.12) during their best periods of fairness. They have it *almost right* for a great majority of their population, the bottom 99%. So it stands to reason that what they have allocated for their top 1% is likely to be their fair share under the *real-life nonideal* conditions.

Sure, it is more than the ideal share allocated in BhuVai, but we don't expect to see that in the real world anyway. The question is: How close can we get to it pragmatically? We believe the Scandinavian numbers are those practically feasible numbers for the top 1%. As we see from the figures, even the United States was in this range, during 1945 to 1975 although near the top end. These trends offer us a valuable practical insight that ∼50% to 100% above the one-class model's ideal value is perhaps the target to strive for to restore a sense of fairness in practice for the top 1%. We can fix this at the beginning, i.e., where the income originates in corporations, or at the end, in after-tax and transfer policies, or a combination of both.

## 6.6  Gini Coefficient versus $\psi$

A common measure used to quantify inequality is the *Gini coefficient*, which ranges from 0 (when everyone gets the same income) to 1 (when all income goes to a single individual, considered to be the most unfair outcome). While the Gini index measures inequality, it does not measure the level of fairness (or unfairness) in inequality. As we have stated before, equality of income is *not* the fairest distribution, as different people contribute differently, whether in a corporation or a society. Therefore, while lower Gini coefficient values generally signal lower inequality, they don't necessarily imply that they also signal the fairness of a society.

Since a lognormal distribution is the fairest outcome, for the sake of comparison, we computed its Gini coefficient, which would make it the ideal value that a country should achieve, for the 12 countries (see figures 6.13a and b). The Gini coefficient for a lognormal distribution is given by:

$$\text{Gini coefficient} = 2\Phi\left(\frac{\sigma}{\sqrt{2}}\right) - 1 \qquad (6.4)$$

where $\sigma$ is the lognormal standard deviation and $\Phi$ is the cumulative density function for the standard normal distribution.

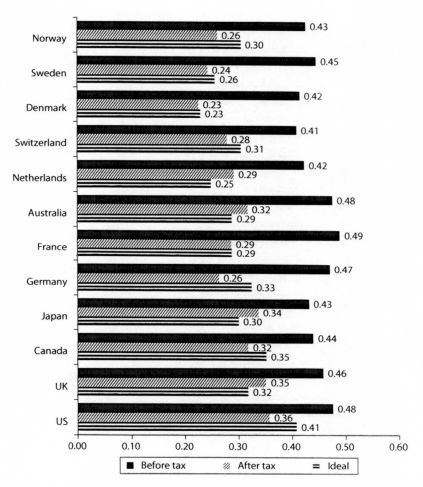

*Figure 6.13a* Gini coefficients calculated from the Organisation for Economic Co-operation and Development data (early to mid-2000s). Source: Organisation for Economic Cooperation and Development (OECD 2015).

The empirical values are obtained from the Organisation for Economic Co-operation and Development (OECD) and the Luxembourg Income Studies (LIS) database and sources (OECD 2015; LIS Database 2015). They correspond to two different time frames (the exact years are not provided in the sources we used), and they show before and after taxes and transfers. As we have seen in figures 6.1 through 6.12, the general trend of Scandinavian countries closely approaching

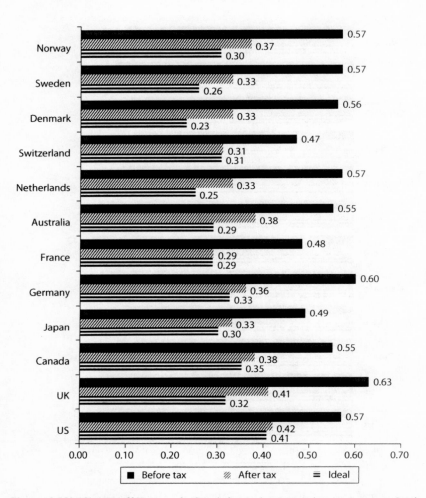

*Figure 6.13b* Gini coefficients calculated from the Luxembourg Income Studies Database data (early to mid-2000s). Source: Luxembourg Income Studies Database (2015).

the ideal distribution is found here as well. Their actual coefficients (after taxes and transfers) are very close to the ideal values, and in some cases they appear to have even overcompensated and dropped below them, according to one set of data (e.g., Norway, 0.26 versus 0.30; Sweden, 0.24 versus 0.26; Switzerland, 0.28 versus 0.31) but not according to the other set (e.g., Norway, 0.37 versus 0.30; Sweden, 0.33 versus 0.26, Switzerland, 0.31 versus 0.31). This is even the case for the

United States (0.36 versus 0.40), which is not equitable, as we see in figures 6.1 through 6.12. Switzerland seems to have overcompensated for fairness according to both data sets. Given our reservations about the Gini coefficient as a measure of equity and fairness, we don't take these discrepancies too seriously and we show the comparison only for the sake of completeness.

Despite our reservations about the Gini coefficient, there is one valuable lesson to be drawn here. The before- and after-tax and transfer values in the Gini coefficients show how macroeconomic policies can be used to achieve greater fairness, a level that approaches ideality, in practice. In that regard, it would be valuable to compare the posttax and transfer income distributions for the three segments with the model predictions (as we did in figures 6.1 through 6.12 for the pretax income). Since we now know what the target (i.e., fairest) distributions are, we could *rationally* design and fine-tune tax and transfer policies that result in the desired, near-ideal, after-tax income distribution for a given society.

For example, in the United States, the ideal income share of the top 1% is 5.8% whereas the actual share is 17.5%. Based on this information, we can design tax rates and income thresholds in such a way that the after-tax income share of the top 1% is closer to ideality (i.e., after-tax $\psi_1$ is closer to zero), thus improving fairness in the society. Similarly for other levels, one can do this for every percentile of the population, all the way down to the bottom. Our theory provides the moral justification for taxation, based on the foundational principle of equality of all citizens and their fair treatment. We, of course, realize that this kind of approach is perhaps more likely to be adopted in Scandinavia, and in other European countries, first before it gets any chance in the United States, given its political climate.

We could also develop similar guidelines for deciding executive compensation in corporations. For instance, in November 2013, Swiss voters considered and rejected a referendum (Garofalo 2013) that would have capped the CEO pay ratio at 1 : 12. The number 12 was decided rather arbitrarily; it was felt that the CEO couldn't make more in one month than what the lowest employee makes in one year. Using our framework, we can examine this more rationally to develop guidelines based on the fundamental principles of economic fairness rather than on arbitrary limits (we discuss this in chapter 7).

Depending on the aggregate data available, one can compute $\psi$ for other segments, such as deciles and quintiles. We can then compute an overall composite coefficient $\Psi$, e.g., by calculating

$$\Psi = w_{90}\psi_{90} + w_{10-1}\psi_{10-1} + w_1\psi_1 \qquad (6.5)$$

where $w$ may be equally weighted, population-weighted, or income-weighted. We have not done this because it is not clear, a priori, what weights would reflect the overall level of fairness (or unfairness) correctly. Careful studies are needed before any recommendation can be made along these lines. Another property of the lognormal distribution that can be used for comparison is, of course, entropy itself, which is given by

$$\frac{1}{2} + \frac{1}{2}\ln(2\pi\sigma^2) + \mu \qquad (6.6)$$

One can then calculate $\psi_{en}$ = actual entropy/ideal entropy − 100%. We can also develop a similar coefficient by using the Theil index instead of entropy, because they are, of course, closely related. Again, careful further studies are needed to identify the most useful nonideal inequality coefficient. However, it is clear that we need an appropriate reference to compute that, and it is our proposal that we use the lognormal distribution as that ideal basis.

## 6.7 One-Class and Two-Class Societies: Simulation Results

To test our theory's predictions regarding multiclass societies, we ran an agent-based simulation comprising 1 million agents in two classes, at 100 salary levels, with a minimum pay of $20,000 (using a minimum wage of $10 per hour and 2000 hours per year) and a maximum pay of $3,000,000. We chose a two-class society because empirical income data from different countries seem to suggest that the bottom 95% to 97% of a population follows a lognormal distribution, while the top 3% to 5% follows a power law or Pareto distribution (Champernowne and Cowell 1998; Chatterjee et al. 2005; Chakrabarti et al. 2013). So we explored the typical case of 95% of the population in class 1 and 5% in class 2. The respective $\alpha$, $\beta$, and $\gamma$ values for the two classes are shown in table 6.2. The dynamics unfolds by each agent trying to maximize

TABLE 6.2
Two-Class System Parameters

| $j$ | $\alpha_j$ | $\beta_j$ | $\gamma_j$ |
|---|---|---|---|
| 1 | 93.4 | 3.87 | 2.17 |
| 2 | 95.8 | 3.67 | 4.34 |

its utility given by equation (3.6) by switching from its current job to a better one, and the equilibrium (stationary) distribution emerges over time, as shown in figure 6.14.

In figure 6.14, the gray (class 1) and black (class 2) histogram bars are data from the simulation, and the lines are predictions by the model. As the results show, the two populations are sufficiently separated; hence the individual lognormal distributions predicted by the model [equation (5.50)] fit the data very well. For the population shown in black (class 2), its higher $\alpha$ makes it value the utility from salary more, its lower $\beta$ motivates it to put in greater effort, and its higher $\gamma$ makes the utility from future prospects more important, compared to the class 1 agents. As a result, class 2 agents are averse to jobs with lower pay. It is the opposite for the agents from class 1. Hence they separate, almost like phase separation in physicochemical systems, or the separation of

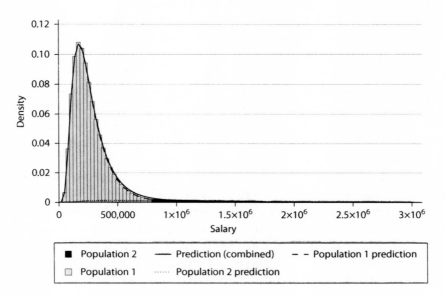

*Figure 6.14* Agent-based simulation results.

oil and water in a mixture. We observe that the combined distribution (solid black line), as one might expect, fits the lognormal distribution for the gray population (class 1) quite well in the lower and medium salary ranges, but deviates from it for higher salaries.

We also show that the distribution of the higher salaries (largely occupied by the black population agents) can be fitted to an inverse power law, given as

$$x_i \propto S_i^{-(1+\eta)} \qquad (6.7)$$

1. Top 3% fitted: $\eta = 1.60$, $r^2 = 0.99$
2. Top 5% fitted: $\eta = 1.70$, $r^2 = 0.96$

We see that the inverse power law fit is very good for both the top 3% and to 5%. The Pareto exponents from our simulation data agree well with empirical data reported in the literature—between 1 and 2, but typically around 1.5 for the top 3% (Chakrabarti et al. 2013). Thus, the main lesson here is that while the overall distribution is a combination of two lognormal distributions, it can be quite easily misidentified as a lognormal for the majority and an inverse power law or Pareto distribution for the minority at the top end of the salaries. This again confirms similar warnings by (Perline 2005) and (Mitzenmacher 2004). For actual salary distributions reported in the literature (Chakrabarti et al. 2013), the available data are not good enough to sort this out clearly, and further studies are needed.

Our two-class approach can be generalized for a $\Pi$-class game along the lines we described above. However, as noted, we might need only three classes, at most four, to model empirical data effectively. At any rate, at present the empirical data reported in the literature are not good enough to test three-class or four-class models. The best it seems to be able to do is to identify the need for a two-class model, but even there it appears unable to discriminate between a lognormal distribution and a power law fit for the top 3% to 5% as shown in figure 6.15.

An interesting question is, Why does the two-class split in actual data occur at about 95% to 97% of the population? Why not at 80%, for instance? In our theory, this is related to the fraction of the population that is highly motivated, talented, and driven to individual accomplishments and success (i.e., class 2). It would be nice if we had demographic data that directly showed where this two-class split occurs in the real world, but we don't.

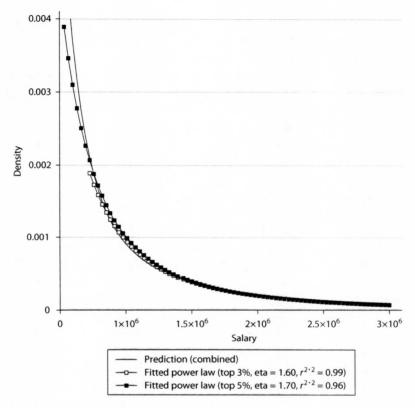

*Figure 6.15* Power law fits truncated lognormal.

## 6.8 Summary

To summarize, we compared the predictions made by our theory with empirical data on pretax income (excluding capital gains) distributions from different countries. We defined a new measure of income inequality $\psi$, which quantifies how fair the inequality is in real-world societies.

So how close are the real-world societies to BhuVai? As we expected, none are ideal, but there are some pleasant surprises. In Scandinavia, the income shares of the bottom 90% and top 10% to 1% are remarkably close to the ideal BhuVai values over the last ~20 years. This is surprising because we didn't expect any real-world society to come this close to BhuVai, given the restrictive assumptions and approximations in the ideal model. For an overwhelming majority of the population (~99%), Norway, Sweden, Denmark, and Switzerland (and to a lesser extent, the Netherlands and Australia) have achieved income shares that are quite close to the values for the ideal free market. It is quite remarkable that these are before taxes and transfers.

What is even more surprising is that these societies did not know, *a priori*, what the ideal, theoretically fairest, distribution was, and yet they seem to have "discovered" a near-ideal outcome empirically on their own and have maintained it over 20 years.

The United Kingdom and the United States are at the other end of the inequality spectrum, with a high degree of unfairness, which is not a surprise.

Based on our analysis, one could argue that the aforementioned countries are functioning almost like the free-market societies envisioned by Adam Smith. Through an enlightened combination of individual, corporate, and societal values, and macroeconomic policies, these countries have made sure that the interests of the bottom $\sim$90% are not trampled upon by the rent-seeking market power of the top $\sim$10%.

It's instructive to compare Scandinavia and the USSR. In both cases, there is state intervention in the economy. Both are socialistic to varying degrees. However, while the state intervention in Scandinavia somehow managed to produce a near-ideal free-market outcome, the centrally planned USSR economy was a total disaster. There are lessons to be learned here—it appears there is a role for optimal state participation.

Real-life free markets are not level playgrounds, where rent seeking is widespread, market power is routinely exploited, and the influence of the working class is steadily diminishing. As we discussed in section 3.1, the free market, ideal or otherwise, itself is a human creation, and, as Reich (2015) observes, "Few ideas have more profoundly poisoned the minds of more people than the notion of a 'free market' existing somewhere in the universe, into which government 'intrudes.'... Government doesn't 'intrude' on the 'free market.' It creates the market." In Scandinavian societies, it seems that the government has defined the "rules of the game" in such a way that the resulting free market behaves almost ideally.

These countries, particularly Scandinavia, are generally derided by free-market fundamentalists as socialist welfare states, while the United States is generally considered to be the shining example of free-market capitalism. Our theory suggests, in the context of distributive justice, that it is quite the contrary. The United States is where the free-market mechanism has broken down the most. While the American free-market democracy has many great things going for it—as evidenced by its ability to attract copious global talent and the constant stream of innovations produced year after year, leading to economic growth and dominance—we have dropped the ball on distributive justice in the last three decades.

Since $\psi$ identifies the target (i.e., fairest inequality) quantitatively, it provides quantitative guidance for formulating tax and transfer policies that can correct for the inequities in pretax income such that after-tax $\psi$ is close to zero, thus restoring fairness in the society.

We conclude with a comparison of our predictions for a two-class society with the results from agent-based simulations and empirical data, which support our claims. We observe that while the overall distribution is a combination of two lognormal distributions, it can be quite easily misidentified as a lognormal for the majority (bottom 95-97%) and an inverse power law or Pareto distribution for the minority at the top (3-5%).

It is important to highlight an intriguing outcome of our analysis. Observe that properties like utility, welfare, value etc., are difficult to measure and quantify in practice. This, and their subjective nature, have long been a source of concern and criticism in economics. Unlike energy, the source of inspiration for the neoclassical models, which can be measured precisely, utility or value cannot be measured. However, even though our theory is built on the non-measurable quantity utility, it finally yields a result—the income distribution—that can be measured, quantified, and verified empirically. In that sense, this is like the wave function in quantum mechanics, a complex quantity which in itself is not measurable but facilitates the prediction of other measurable quantities. We are not, of course, suggesting that utility is a complex quantity. We only observe that even non-measurable quantities can lead to measurable and useful results at the end. The empirical verification of our theory's prediction of lognormal income distribution can be seen as a verification of the logarithmic utility assumption, $u = \ln S$.[2] This implies that this directly non-measurable quantity can now be measured indirectly.

We now shift our focus from country-wide income distributions to a company specific topic of great importance, namely, pay distribution and executive compensation in companies. Can our theory provide some guidance in this matter? As it turns out, it can, and this is what we address in chapter 7.

CHAPTER SEVEN

# What Is Fair Pay for Executives?

> In judging whether Corporate America is serious about reforming itself, CEO pay remains the acid test. To date, the results aren't encouraging.
>
> WARREN BUFFETT

> The poor have sometimes objected to being governed badly. The rich have always objected to being governed at all.
>
> G. K. CHESTERTON

One of the most important concerns in corporate governance of our times is the runaway compensation packages for corporate executives. As the quote from Warren Buffet reminds us, this remains the "acid test." "The current levels of compensation for CEOs in corporate America are, in a word, outrageous," said Jack Bogle (Bloxham 2011), founder of the Vanguard Group, who is unlikely to be mistaken for a communist or a socialist. The ratio of CEO salary (i.e. total compensation including bonuses and stock options) to that of an average employee has risen from about 25 to 40 in the 1970s to over 300 in recent years in the United States. Furthermore, as Reich (2015, 98) observes, "The share of corporate income devoted to compensating the five highest-paid executives of large public firms went from an average of 5% in 1993 to more than 15% in 2013."

This naturally raises the question: What is fair pay for a CEO (and other senior executives) relative to other employees?

A common response to this question is that the free market determines the value of a CEO and the other employees. Is the market really effective in determining CEO pay? This is perhaps true in an ideal free market, where there is no rent seeking or managerial power exploitation by the executives. But as we discussed in chapter 6, real-life free markets are far from ideality in many countries, particularly in the United States. We recall Robert Reich (2015, xiv): "The meritocratic claim that people are paid what they are worth in the market is a tautology that begs the questions of how the market is organized and whether that organization is morally and economically defensible. In truth, income and wealth increasingly depend on who has the power to set the rules of the game." This vital point is also emphasized by Bebchuk and Fried (2006), who have documented how managerial power has played a key role in sharply increasing executive pay. Herb Simon, Nobel Laureate in Economics, once remarked (Reich 2015, 91): "If we are generous with ourselves, I suppose we might claim that we 'earned' as much as one-fifth of our income. The rest is the patrimony associated with being a member of an enormously productive social system."

So if the free market were to behave ideally, what would the CEO pay be like? Our theory provides guidance in answering this question. As we saw, the fairest distribution of pay in a large corporation functioning in an ideal free-market environment is the lognormal distribution. Again, we must caution that our theory is not valid for small, highly entrepreneurial organizations where a handful of employees (e.g., the founders of a start-up) are demonstrably much more valuable than the others. But this is hardly the case for many large organizations (e.g., Fortune 500 companies) with tens to hundreds of thousands of employees, where the CEO is often another "hired hand," as J. P. Morgan once remarked (Drucker and Schlender 2004).

How does our prediction compare with reality? The pay distribution data for individual companies are generally not available and are often regarded as proprietary information. However, one can compare with aggregate data reported in the literature gathered from income tax filings for millions of people. For large corporations that employ tens of thousands of people (e.g., Fortune 500 companies), it is reasonable to expect the essential character of the pay distribution for such corporations to be quite close to that of the population data. The mean and variance might differ, but it is reasonable to expect the essential qualitative nature to remain the same.

Our prediction of a lognormal distribution is in good agreement with the population data (Champernowne 1953; Champernowne and

Cowell 1998; Pareto 1897; Gini 1921; Montroll and Shlesinger 1982; Chatterjee et al. 2005, 2007) for the bottom part of the spectrum. As discussed in chapter 6, it's been reported empirically that typically the bottom 95% to 97% of the income distribution follows the lognormal distribution while the top 3% to 5% follows a Pareto, or power law, distribution. However, as we discussed, we believe the Pareto tail could very well be the tail of another lognormal distribution (different from the one that models the bottom 95% to 97%) that has been misidentified. This issue needs to be sorted out with more careful empirical studies. Furthermore, the general agreement on the overall functional form does not necessarily imply that the income shares by the different percentiles of the population are in agreement, as we discussed at length in chapter 6.

## 7.1 Theory's Predictions for Large Corporations

Ideally, top management's pay also should fall in line with the rest of the employee's pay on a lognormal distribution. Otherwise, we are treating the contributions of the bottom 95% to 97% of the employees unfairly, by overpaying for the services of top management. Further, the company is also wasting substantial resources by overcompensating senior management (Bebchuk and Fried 2005). Then there is also the risk of creating poor morale among the vast majority of the employees (Bebchuk and Fried 2005; Drucker and Schlender 2004; Lazear and Shaw 2007).

So how should a compensation committee go about determining a fair pay structure for top management? One procedure is to best-fit the salary data to a lognormal distribution for the bottom 95% to 97% of employees and estimate the mean and variance. This is the distribution that the rest of the data (i.e., top 3% to 5%) ought to follow, so that the entire population of employees is treated fairly on the same basis. The CEO's salary can be determined directly from this extrapolated lognormal distribution plot or by calculating the standard deviation for the CEO's data position ($\sigma_{CEO}$) on the bell curve and computing the corresponding salary for it. Adjustments to this ideal pay package can be made as needed, to account for individual talents and accomplishments, by relating them to both short-term and long-term performance targets.

Since the salary distribution data for companies are not available, we present as an example (table 7.1) a number of plausible salary scenarios, and we estimate what fair pay for a CEO ought to be for the top CEOs whose 2008 pay packages were published by the *New York Times* (Jones 2009a).[1] We consider 35 companies from the top 50 for which reliable employee data were available from various websites. Since ln $S$ follows a normal distribution, we perform our calculations using the standard normal distribution tables, which make the calculations easier. In our calculations, we assume that

$$\ln(\text{mean salary}) - \ln(\text{minimum salary}) = 3\sigma \qquad (7.1)$$

While this is not a very accurate approximation, this is the best we can do in the absence of company-specific income data. If such data were available—for instance, such data would be available to the compensation committee of a corporation—there are much better statistical procedures available to estimate the mean and standard deviations more accurately. However, our current estimate from equation (7.1) gives a general sense of the CEO pay ratio, which is adequate for the present discussion.

From this we estimate $\sigma$ and calculate the CEO salary by computing the corresponding $\sigma_{CEO}$ upper bound for which the outside area under the bell curve (i.e., under the far right tail) equals $1/N$, where $N$ is the total number of employees in the company. For example, for Motorola which had about 64,000 employees (as of 2008), $1/64,000 = 0.0000156$ corresponds to a position on the bell curve that is 4.16 standard deviations (i.e., $\sigma_{CEO} = 4.16\sigma$) to the right of the mean, thereby fixing the ideal CEO salary. We emphasize, again, that our computations are only rough estimates, as we lack detailed salary distribution data, which are needed to estimate the $\mu$, $\sigma$, and $\sigma_{CEO}$ more accurately. Once we have the actual pay distribution, we can apply much more rigorous statistical methodologies to better estimate these parameters and the CEO pay.

Nevertheless, even these rough estimates are quite illuminating, as shown in table 7.1 (based on 2008 data). The minimum salary is assumed to be $25,000 or $25K per year. We consider several average annual-salary scenarios: $40K, $60K, $80K, and $100K. The CEO pay ratios are computed by dividing the actual, and the ideal, annual CEO pay by the minimum annual salary. For scenario 2, which is perhaps

TABLE 7.1
Comparing Top CEO Salaries with Minimum Annual Salary of $25K: Actual versus Ideal Ratios

| Company | CEO's total pay incl. salary + bonus in $ millions 2008 | Total # of employees estimates from websites 2008 | Area outside (1/N) from normal distribution z-tables | CEO sigma estimate from z-tables | Minimum salary | Actual CEO pay ratio B/F | Mean salary scenario 1 | Ideal CEO pay ratio | Mean salary scenario 2 | Ideal CEO pay ratio | Mean salary scenario 3 | Ideal CEO pay ratio | Mean salary scenario 4 | Ideal CEO pay ratio |
|---|---|---|---|---|---|---|---|---|---|---|---|---|---|---|
| Motorola | 104.4 | 64,000 | 0.00001563 | 4.16 | 25,000 | 4,176 | 40,000 | 3.1 | 60,000 | 8.1 | 80,000 | 16.0 | 100,000 | 27.3 |
| Oracle | 84.6 | 50,000 | 0.00002000 | 4.11 | 25,000 | 3,384 | 40,000 | 3.0 | 60,000 | 7.9 | 80,000 | 15.7 | 100,000 | 26.6 |
| Walt Disney | 51.1 | 150,000 | 0.00000667 | 4.35 | 25,000 | 2,044 | 40,000 | 3.2 | 60,000 | 8.5 | 80,000 | 17.3 | 100,000 | 29.9 |
| American Express | 42.8 | 30,162 | 0.00003315 | 3.99 | 25,000 | 1,712 | 40,000 | 3.0 | 60,000 | 7.7 | 80,000 | 15.0 | 100,000 | 25.2 |
| Citi Group | 38.2 | 324,850 | 0.00000308 | 4.52 | 25,000 | 1,528 | 40,000 | 3.2 | 60,000 | 9.0 | 80,000 | 18.4 | 100,000 | 32.2 |
| HP | 34.0 | 321,000 | 0.00000312 | 4.52 | 25,000 | 1,360 | 40,000 | 3.2 | 60,000 | 9.0 | 80,000 | 18.4 | 100,000 | 32.2 |
| News Corp | 30.1 | 53,000 | 0.00001887 | 4.12 | 25,000 | 1,204 | 40,000 | 3.0 | 60,000 | 8.0 | 80,000 | 15.8 | 100,000 | 26.8 |
| Honeywell | 28.7 | 128,000 | 0.00000781 | 4.32 | 25,000 | 1,148 | 40,000 | 3.1 | 60,000 | 8.5 | 80,000 | 17.1 | 100,000 | 29.4 |
| P&G | 25.6 | 138,000 | 0.00000725 | 4.34 | 25,000 | 1,024 | 40,000 | 3.2 | 60,000 | 8.5 | 80,000 | 17.2 | 100,000 | 29.7 |
| Abbott | 25.1 | 68,697 | 0.00001456 | 4.18 | 25,000 | 1,004 | 40,000 | 3.1 | 60,000 | 8.1 | 80,000 | 16.2 | 100,000 | 27.5 |
| Lockheed Martin | 22.9 | 146,000 | 0.00000685 | 4.35 | 25,000 | 916 | 40,000 | 3.2 | 60,000 | 8.5 | 80,000 | 17.3 | 100,000 | 29.8 |
| eBay | 22.5 | 7,769 | 0.00012872 | 3.65 | 25,000 | 900 | 40,000 | 2.8 | 60,000 | 7.0 | 80,000 | 13.2 | 100,000 | 21.6 |
| Anadarko Petroleum | 22.2 | 4,000 | 0.00025000 | 3.48 | 25,000 | 888 | 40,000 | 2.8 | 60,000 | 6.6 | 80,000 | 12.3 | 100,000 | 19.9 |
| United Tech | 22.0 | 223,100 | 0.00000448 | 4.44 | 25,000 | 880 | 40,000 | 3.2 | 60,000 | 8.8 | 80,000 | 17.9 | 100,000 | 31.1 |
| BMS | 21.8 | 42,000 | 0.00002381 | 4.07 | 25,000 | 872 | 40,000 | 3.0 | 60,000 | 7.9 | 80,000 | 15.5 | 100,000 | 26.2 |
| Hess | 21.3 | 13,300 | 0.00007519 | 3.79 | 25,000 | 852 | 40,000 | 2.9 | 60,000 | 7.2 | 80,000 | 13.9 | 100,000 | 23.0 |
| Johnson and Johnson | 21.1 | 118,700 | 0.00000842 | 4.30 | 25,000 | 844 | 40,000 | 3.1 | 60,000 | 8.4 | 80,000 | 16.9 | 100,000 | 29.1 |
| IBM | 21.0 | 398,455 | 0.00000251 | 4.57 | 25,000 | 840 | 40,000 | 3.3 | 60,000 | 9.1 | 80,000 | 18.8 | 100,000 | 33.0 |
| Verizon | 19.9 | 234,971 | 0.00000426 | 4.45 | 25,000 | 796 | 40,000 | 3.2 | 60,000 | 8.8 | 80,000 | 17.9 | 100,000 | 31.2 |
| Coca-Cola | 19.6 | 90,500 | 0.00001105 | 4.24 | 25,000 | 784 | 40,000 | 3.1 | 60,000 | 8.3 | 80,000 | 16.5 | 100,000 | 28.3 |
| Avon | 19.5 | 42,000 | 0.00002381 | 4.07 | 25,000 | 780 | 40,000 | 3.0 | 60,000 | 7.9 | 80,000 | 15.5 | 100,000 | 26.2 |
| Cisco | 18.8 | 32,160 | 0.00003109 | 4.01 | 25,000 | 752 | 40,000 | 3.0 | 60,000 | 7.7 | 80,000 | 15.1 | 100,000 | 25.5 |
| Qualcomm | 18.6 | 11,932 | 0.00008381 | 3.76 | 25,000 | 744 | 40,000 | 2.9 | 60,000 | 7.2 | 80,000 | 13.7 | 100,000 | 22.7 |
| General Dynamics | 18.0 | 83,500 | 0.00001198 | 4.22 | 25,000 | 720 | 40,000 | 3.1 | 60,000 | 8.2 | 80,000 | 16.4 | 100,000 | 28.1 |
| CVS | 17.4 | 160,000 | 0.00000625 | 4.37 | 25,000 | 696 | 40,000 | 3.2 | 60,000 | 8.6 | 80,000 | 17.4 | 100,000 | 30.1 |
| Merck | 17.3 | 58,900 | 0.00001698 | 4.15 | 25,000 | 692 | 40,000 | 3.1 | 60,000 | 8.0 | 80,000 | 16.0 | 100,000 | 27.2 |
| Prudential | 16.3 | 49,616 | 0.00002015 | 4.11 | 25,000 | 652 | 40,000 | 3.0 | 60,000 | 8.0 | 80,000 | 15.7 | 100,000 | 26.7 |
| Deere | 16.2 | 52,022 | 0.00001922 | 4.12 | 25,000 | 648 | 40,000 | 3.0 | 60,000 | 8.0 | 80,000 | 15.8 | 100,000 | 26.8 |
| AT&T | 15.0 | 302,660 | 0.00000330 | 4.51 | 25,000 | 600 | 40,000 | 3.2 | 60,000 | 8.9 | 80,000 | 18.4 | 100,000 | 32.1 |
| ADM | 15.0 | 27,600 | 0.00003623 | 3.97 | 25,000 | 600 | 40,000 | 3.0 | 60,000 | 7.6 | 80,000 | 14.9 | 100,000 | 25.0 |
| Pepsi | 14.9 | 198,000 | 0.00000505 | 4.41 | 25,000 | 596 | 40,000 | 3.2 | 60,000 | 8.7 | 80,000 | 17.7 | 100,000 | 30.6 |
| Johnson Controls | 14.9 | 140,000 | 0.00000714 | 4.34 | 25,000 | 596 | 40,000 | 3.2 | 60,000 | 8.5 | 80,000 | 17.2 | 100,000 | 29.7 |
| Pfizer | 14.8 | 86,600 | 0.00001155 | 4.22 | 25,000 | 592 | 40,000 | 3.1 | 60,000 | 8.2 | 80,000 | 16.4 | 100,000 | 28.1 |
| Boeing | 14.8 | 162,200 | 0.00000617 | 4.37 | 25,000 | 592 | 40,000 | 3.2 | 60,000 | 8.6 | 80,000 | 17.4 | 100,000 | 30.1 |
| Burlington | 14.6 | 40,000 | 0.00002500 | 4.06 | 25,000 | 584 | 40,000 | 3.0 | 60,000 | 7.8 | 80,000 | 15.4 | 100,000 | 26.1 |
| Average | 26.4 | | | | | 1,057 | | 3.1 | | 8.2 | | 16.3 | | 27.9 |

closer to reality, the ideal pay ratio is 8.2. On average, the actual CEO pay ratio was 1,057 (based on the top 35 companies' data in 2008), while the ideal value is about 8.2, or about 129 times the ideal value. In 2008, S&P 500 CEOs averaged about $10.0 million, with an average CEO pay ratio of about 400 (on the $25,000 per year minimum salary basis) which is about 50 times the ideal value.

Recall the actual CEO pay ratio chart in figure 1.6 and the CEO pay ratio in table 1.1., from chapter 1. Using the ratio of 354 for the United States, it is about 43 times the ideal value, similar to the 2008 figure. It's instructive to compare the ideal ratio predicted by our theory with those of what people in the survey thought it should be (table 1.1). People from Australia, France, Germany, Japan, and the United States agreed on a ratio of around 6 to 8, which is close to our theoretical value. Also note from the chart that in the 1960s and 1970s, the ratio was 20 to 40 in the United States, which was only 3 to 5 times the ideal value.

Obviously, we do expect real-world economic systems to deviate from ideal systems, thus necessitating larger pay packages for CEOs. But would they deviate so much that the actual CEO pay ratio is 40 to 50 times the ideal benchmark? Something like 3 to 5 times is more reasonable (i.e., a ratio of 20 to 40), which is what it was in the 1960s and 1970s. Note that this is more in line with what people estimated the present-day ratio to be (in middle column table 1.1).

It is also instructive to note that, in 2006, according to the *Wall Street Journal* (Etter 2006), the average CEO pay ratio was about 11:1 in Japan, 15:1 in France, 20:1 in Canada, and 22:1 in Britain. These figures are not that far off (compared to the U.S. ratios) from the ideal benchmark estimates. As noted, even in the United States the CEO pay ratios were much more reasonable and in general agreement with the ideal values, in the 1960s and early 1970s. Thus, the executive pay excesses appear to be a phenomenon of the past two decades, the new gilded age, in the United States, perhaps due to the reasons argued by Bebchuk and Fried (2005), Reich (2015), and Stiglitz (2015). This appears to be another valuation bubble—the CEO valuation bubble, much like the ones we have witnessed in stocks, real estate, commodities, etc. While the emphasis of this section has been on CEO pay, the observations made here are applicable to all senior management in a corporation.

## 7.2 How Much of a Company's Success Is due to the Executives?

Given the amount of press coverage and the extent of the shareholder and public outrage in recent years, one would have expected swifter corrective actions on the excessive CEO pay front; but as Buffett himself has lamented, not much progress has been made so far. So it is worth hammering this issue again, now that we have a theoretical framework to discuss the fairness of executive compensation.

So how much of a company's success is due to the executives? This difficult question is at the heart of the executive compensation and the income inequality debate. In a large corporation, all employees contribute to the company's overall success in their own ways. Take, for example, Walmart, which employed about 2.2 million workers in 2014; or IBM, which employed about 412,000; or GE's 300,000, Coca-Cola's 123,000, and ExxonMobil's 75,000. The average number of employees was about 54,000 for the Fortune 500 companies in 2014 (Zillman and Jones 2015). Every one of them contributed to the overall success of the company she worked for—admittedly, some more, others less.

But how much of the company's success is due to the efforts of the CEO and the top executives? How much is due to the other workers?

The honest answer is that nobody really knows for certain, in most cases, as we discuss below.

### 7.2.1 Are CEOs like Sports and Movie Stars?

Some have compared CEOs with star athletes, movie stars, and other leading figures in the entertainment industry, to justify their extraordinary compensation (Bares 2013; Hallock 2012a; Fernando 2009). Apparently, some economists are puzzled that while there is so much resentment about excessive CEO pay, there are no such ill feelings toward the extraordinary incomes of sports stars and movie legends (Hallock 2012b). The "CEO as a sports/movie star" line of defense seems plausible superficially, but is wrong. So let us examine this from the perspective of our theory.

In the case of film stars, sports legends, and others in the entertainment industry, certain individuals are *demonstrably much superior* to their colleagues. For instance, for a Michael Jordan, a Sachin Tendulkar, and a David Beckham, one can reliably measure their accomplishments in terms of a variety of statistics regarding their baskets, runs, goals, etc. scored over many games and over many years. But that is hardly the case for most corporate CEOs and other executives. They don't have such measurable and reliable individual performance metrics, validated "game" after "game," publicly witnessed by millions of customers. A lot of their achievements are not individual achievements, but team-based accomplishments. Yes, victories in sports and movies are also team-based, but the contributions of the different individuals, in scoring or acting, are often quite clear for everyone to see.

When one goes to see a *Mission Impossible* movie, one knows that Tom Cruise is the main action hero. Or when people go to see a Tom Hanks or Meryl Streep movie, sometimes they may not even know the movie title or who else is in it. They are going to see a Tom Hanks movie or a Meryl Streep movie. When we read and enjoy the Harry Potter series and admire its resounding success, it is quite clear as who should get the lion's share of the credit. It is J. K. Rowling, of course! Most people, including this author, have no problem, whatsoever, with her financial rewards, or of others such as the ones mentioned above. These people are the main attraction, and their individual contributions are quite evident for everyone to see.

But that's hardly the case when one buys a GM car or a Coke product. Who knows, and more importantly who cares, who the CEO is? You are not buying the product because so-and-so is the CEO. But that is not the case with Cruise, Hanks, Streep, Spielberg, Jordan, Tendulkar, or Rowling. They *are* the products! You are "buying" their performance for the duration of the event—more accurately, you are "buying" the enjoyment derived from their performances.

Again, that is not the case with many corporate CEOs and other senior executives. Their success is built on the backs of tens to hundreds of thousands of their employees. When their companies succeed, it is not at all clear how much of this success should be credited to the efforts of the CEO and the top executives, and how much should be credited to the thousands of workers. On the other hand, it is quite evident when Jordan, Tendulkar, or Beckham plays, or Cruise, Hanks, or Streep acts, or Rowling writes, how much of the success of their enterprises is due to their individual contributions.

Even in the case of sports and movies, others such as coaches and managers also contributed to the overall success in important ways. So even in these businesses where one has copious statistics for individual accomplishments, it gets to be challenging to assign credit fairly. For other businesses, such as Fortune 500 companies, where such reliable and verifiable metrics are largely absent, it gets to be downright murky. The metrics often used by the compensation committee, and external compensation consultants, to justify a CEO's compensation are just phrenology at best for most large corporations. For instance, one metric frequently used is the performance of the company stock. But how much of the appreciation is due to the underlying bull trend of the market? Even if the CEO had done nothing special, the stock is likely to have appreciated anyway. As the Wall Street adage goes, "Never confuse genius with a bull market!"

### 7.2.2 Top Five Executives Grabbing 15% of the Income Pie

However, this confusion and hero worship, as Drucker observes, seem to be rampant in recent years. In general, senior management's contributions are typically overvalued by friendly boards, compensation committees, and consultants; hence they are generally overpaid, particularly in the United States. Reich (2015, 98) said, "The share of corporate income devoted to compensating the five highest-paid executives of large public firms went from an average of 5% in 1993 to more than 15% in 2013." One finds it hard to believe that just five people, out of a typical workforce of 50,000 employees in a large public corporation, added 15% of the total value, year after year. And that they were adding only 5% of the value just twenty years back.

Are we to believe that the new-generation CEOs are such supermen and superwomen that they were able to triple their added value in just twenty years? As noted, the CEO pay ratio lately is around 300 while it was about 30 in the 1960s, in the United States. Again, are we to believe that the current CEOs are contributing ten times the performance of their predecessors? There may be exceptions, as noted below, but to expect us to believe that such a Herculean feat is accomplished in company after company, year after year, stretches incredulity. A more credible explanation is that this is the result of sheer greed, rent seeking, and broken down corporate governance, as Bebchuk and Fried, Drucker, Crystal, Reich, Stiglitz, and others have documented conclusively.

It is important to recognize that there are exceptions here. As stated, Steve Jobs, who was synonymous with Apple, comes to mind. Warren Buffet would be another such exception. There are a few others like that. But interestingly Jobs and Buffett both earned ordinary compensation, and their wealth came from the ownership of their companies. Many of these exceptional individuals are (or were) active founders (or cofounders), who remain (or remained) with the outstanding companies they created for a long time. Most Americans, including this author, would have no objection to wealth being accumulated in this manner.

But this is not true for many CEOs. As J. P. Morgan observed, "The CEO is just a hired hand." And Morgan, who has never been mistaken as a communist or a socialist, one might add, recommended that the CEO pay ratio should be 20:1, which is in line with our theory's recommendation. How many CEOs are significant long-term owners of the companies they serve? Jobs's annual salary was just $1, but he owned several million shares of Apple, the company he cofounded. When the shares rose, he became wealthy, and that incentive worked out pretty well for the other shareholders and employees as well. Apple's stock price rose from a low of about $3 in 1997, when Jobs returned as CEO, to about $365 when he retired in 2011. How many of the excessively paid CEOs in the United States can claim such a record?

While one cause of the extreme income inequality is the runaway compensation of C-suite executives, the other is the depressed income of the rank and file workers, the bottom 90%. An often cited reason for their income stagnation is the outsourcing of their jobs due to globalization to cut costs. As we outsource low-level jobs in IT and manufacturing to India, China, and other countries, where one could hire such workers at a fraction of their salaries in the United States, there is downward pressure on salaries in the United States for these jobs. While we can understand the economics of this process, we find again that the cost-cutting forces of globalization seem to act selectively, affecting only the rank and file workers, and not the senior management.

More than anything else, it is the lack of accountability and naked greed in the executive suite that get people upset about executive compensation. When rank and file workers mess up, they get fired. When it is the CEO, typically he gets to continue in the job with their multimillion-dollar compensation package. This is not the case when a sports star fails to perform, for the consequences are quite clear and immediate. On the rare occasion that the CEO gets fired, he floats away in his golden parachute, again worth millions of dollars.

Such occurrences are not isolated instances, and they are more frequent than they ought to be in a properly functioning capitalist system. Let us recount some recent examples, as listed by Reich (2015, 104–05):

> Martin Sullivan, who got $47 million when he left AIG, even though the company's share price dropped by 98% on his watch and American taxpayers had to pony up $180 billion just to keep the firm alive; Thomas E. Freston, who lasted just nine months as CEO of Viacom before being fired, and departed with a severance payment of $101 million; Michael Jeffries, CEO of Abercrombie & Fitch, whose company's stock price dropped more than 70% in 2007, but who received $71.8 million in 2008, including a $6 million retention bonus; William D. McGuire, who in 2006 was forced to resign as CEO of UnitedHealth over a stock-options scandal, and for his troubles got a pay package worth $286 million; Hank A. McKinnel, Jr., whose five-year tenure as CEO of Pfizer was marked by a $140 billion drop in Pfizer's stock market value but who left with a payout of nearly $200 million, free lifetime medical coverage, and an annual pension of $6.5 million.... Douglas Ivester of Coca-Cola, who stepped down as CEO in 2000 after a period of stagnant growth and declining earnings, with an exit package worth $120 million; and, as I have noted, Donald Carty, former CEO of American Airlines, who established a secret trust fund to protect his and other executive bonuses even as the firm was sliding into bankruptcy in 2003 and seeking wage concessions from the airline's employees.... The list of shameless CEOs continues to lengthen.

Consider the more recent egregious examples of Turing Pharmaceuticals (Martin Shkreli, CEO) and Mylan NV (Heather Bresch, CEO), another pharmaceutical company, regarding greed, market power abuse, and executive compensation. Turing bought Daraprim, a decades-old drug for treating AIDS patients in 2015 and jacked up the price 5,000%, from $13.50 to $750 per pill. As Alison Kodjak (2016) of NPR reported on the company's internal documents and emails:

When Turing agreed to buy Daraprim, company officials went into celebration mode:

"Very good. Nice work as usual. $1bn here we come." Turing Chairman Ron Tilles's email dated May 27, 2015.

"I think it will be huge. We raised the price from $1,700 per bottle to $75,000.... So 5,000 paying bottles at the new price is $375,000,000—almost all of it is profit and I think we will get 3 years of that or more. Should be a very handsome investment for all of us." Martin Shkreli's email dated August 27, 2015.

"Another $7.2 million. Pow!" Tina Ghorban, senior director of business analytics, reacting to a purchase order for 96 bottles at $75,000 a bottle, on September 17.

Apparently, the public outcry to this abuse was not heard in the boardroom of Mylan when it decided to steeply increase the price of EpiPen (Popken 2016). People with severe allergies often rely on the lifesaving EpiPen when they suffer from potentially fatal allergic attacks. Mylan, which acquired the product in 2007, recently raised the list price for a pair of EpiPen autoinjectors to $600. The price has been rising from a cost of about $100 in 2008. Over the same period, proxy filings show that CEO Heather Bresch's annual total compensation went from $2,453,456 to $18,931,068, a 671% increase.

As Mark Maremont reported in the *Wall Street Journal* (Maremont 2016):

The big pay packages are unusual because of Mylan NV's relatively small size in the U.S. drug industry, where it is No. 11 by revenue and No. 16 by market capitalization.

Mylan's combined total of $292.1 million in pay for its top five executives over the five years ended last December outpaced that of industry rivals several times its size, according to the analysis, including Johnson & Johnson, Pfizer Inc., Bristol-Myers Squibb Co., and Eli Lilly & Co.

Of the 22 companies Mylan named in its 2016 proxy as its preferred peer group for pay purposes—some of them much larger than itself—none paid their top managers more than Mylan over the same five years, the analysis showed.

As we all know, the pharmaceutical companies are not alone in glaringly abusive and greedy practices. It's hard to top Wall Street on this front. As Judge Rakoff (2014), who presided over the subprime financial crisis–related lawsuits, writes in "The Financial Crisis: Why Have No High-Level Executives Been Prosecuted?": "For example, the Financial Crisis Inquiry Commission, in its final report, uses variants of the word fraud no fewer than 157 times in describing what led to the crisis, concluding that there was a systemic breakdown, not just in accountability, but also in ethical behavior." Yet no senior Wall Street executive was prosecuted for his role in the financial crisis.

This patent unfairness, this naked greed, and this widespread lack of accountability among the C-suite occupants are at the root of the resentment felt toward excessive CEO pay. It is not just the size of the enormous pay packages that outrages people, but it is their feeling that the CEOs, and other C-suite occupants, don't deserve them. This also explains why such feelings are absent among the public regarding the fabulous incomes of sports stars, movie actors, and other entertainment figures. People can plainly see them succeed with their individual talents and contributions, thereby deserving what they earn, and people also see that when they fail (sometimes because of unfair practices such as doping), their market disappears along with their big incomes. That's hardly the case with C-suite executives, whose modus operandi seems to be "Heads I win, Tails you lose!"

### 7.2.3   Reining in the Runaway CEO Pay

Clearly, there is considerable concern, even resentment, about the excessive CEO compensation all over the world, and efforts are being made to rein it in. Recent progress on shareholder "say-on-pay" proposals, though nonbinding, is an encouraging sign. For example, in 2013, the Swiss pharmaceutical giant Novartis intended to pay its departing chairman Daniel Vasella a severance package of about $76 million, but the resulting shareholder and public outrage forced Vasella to turn down the package (Thomasson 2013). More recently, according to Kent (2016), the *Wall Street Journal* reported that "BP PLCs shareholders rejected the company's executive pay policy, a stinging—though nonbinding—rebuke to Chief Executive Bob Dudley and his board. Following the company's annual meeting Thursday, the oil giant said 59% of the votes cast were against the company's executive compensation decisions for 2015. That included a roughly 20% increase in Mr. Dudley's total pay for the year, a period during which the company lost $5.2 billion."

In another recent article, the *Wall Street Journal* reported that "With annual meeting season under way, investors across the U.S. can express their anger about high pay for top corporate officers. So-called say-on-pay votes don't give shareholders the power to cut executive rewards. But boards fear a thumbs down on the nonbinding referendum, required since 2011, because the rebuke suggests deeper investor discontent" (Lublin 2016). One gets a pretty good insight into how mad people are when one reads the comments section of these, and other articles on executive pay, reported by the conservative and capitalist bastion, the *Wall Street Journal*. Most readers identify themselves as hard-core capitalists when they complain bitterly about outrageous executive compensation.

Our theory can provide a rational framework for guiding say-on-pay policy discussions. In chapter 6, we mentioned the Swiss referendum on CEO pay in 2013 (Garofalo 2013), where they decided somewhat arbitrarily that the ratio should not be more than 12:1. Rather than deciding arbitrarily, our theory provides a mathematical framework to examine this more rationally and develop guidelines based on fundamental principles of economic fairness. As discussed in sections 5.5.2, 5.5.3, and 7.2.1, unlike sports stars, unbiased individual performance metrics of C-suite executives are generally absent in many large corporations. Under such conditions, the fairness principles guide us toward the lognormal distribution of income as the economically and morally right thing to do. Thus, the lognormal distribution could serve as the *default basis* on which pay adjustments could be made to account for variations on individual performances whenever such metrics are available and justified. Since our theory shows what the ideal shares of the total salary budget should be among the top 1%, next 10% to 1%, and the bottom 90% of employees (or whatever other percentile breakdown one would desire), it can serve as a reference that the compensation committee can tune to accommodate real-world information and constraints.

In this manner, our theory also addresses some thorny issues in pay compression (Lazear and Shaw 2007) by identifying the optimal pay compression policy as a lognormal distribution. Companies may even use our model as a recruiting and retention tool to show how committed they are to a fair treatment of all employees. Shareholders may require companies to publish the income distribution, and maybe even their $\psi$ coefficient, in their annual reports. They could use such information to demand justification for any excess pay paid to top management.

While the overall picture regarding excessive executive compensation seems hopeless, it is encouraging to see instances like the recent announcement by the founder and CEO of Chobani, the yogurt company, Hamdi Ulukaya, who shared his equity up to 10% of the worth of the company among his about 2000 full-time employees. The *New York Times* (Strom 2016) reported that the average payout is about $150,000, with some of the first employees becoming millionaires.

As we have argued, when a company succeeds, nobody knows for certain how much of the success is due to the effort of the CEO and C-suite executives and how much is due to the hard work and dedication of the rest of the employees. Traditional economic theories offer little guidance except to say that the free market sorts this out. As we have shown, this is perhaps true for an ideal free market, but not in the case of actual free markets. As noted, in real life, management's contribution has always been overvalued, while that of the rest is undervalued. So it is indeed encouraging to see some CEOs recognizing this unfairness and making amends.

This message is also articulated by John Mackey and Raj Sisodia (2014) in their book *Conscious Capitalism*, who advocate maximizing *stakeholders'* value rather than mere *shareholders'* value. By stakeholders they mean employees, customers, investors, suppliers, and communities. Mackey, who is a co-CEO and cofounder of Whole Foods Market, also practices what he preaches in his company, which is seen as a model for this new breed of capitalism.

In his foreword to the book, Bill George, the former chairman and CEO of Medtronic, Inc., writes:

> As a committed capitalist, I worry a great deal to see how capitalism has gone off the rails the past quarter century and acquired such a bad name, much of it deserved.
>
> In this book, John Mackey and Raj Sisodia return capitalism to its roots. They make a compelling case for capitalism as the greatest wealth creator the world has ever known. In these pages, they call their version conscious capitalism. I consider it just capitalism, as it is the only authentic form of capitalism. Other forms of doing business, including "crony capitalism," are simply inauthentic versions of the real thing. As we witnessed during the global economic meltdown of

2008 and the Great Recession that followed, these false versions of capitalism cannot be sustained and are doomed to fail over the long term.

Adam Smith would have agreed. As we argued in section 5.10, the long-term stability of a free-market society depends on its robustness and resiliency properties, which are derived from its fundamental principle of fairness. This is what is captured by stakeholder value (i.e., this accounts for both efficiency and fairness) as opposed to shareholder value, which essentially focuses on only efficiency. Hopefully, this nascent trend of conscious capitalism will catch on sooner rather than later.

As noted, we have made simplifying assumptions since our objective is to develop a general theoretical framework and identify general principles that are not restricted by domain-specific details and constraints. Clearly, the next steps are to conduct more comprehensive studies of salary distributions in various organizations, in order to understand in greater detail the deviations from ideality in the marketplace. Agencies such as the Bureau of Labor Statistics and National Bureau of Economic Research could organize task forces to gather wage data from various companies and organizations. The data should be so grouped to analyze wage distribution patterns across several dimensions, such as (1) organization size—small, medium, large, and very large number of employees; (2) different industrial sectors; (3) different types such as private corporations, governments (state and federal), nonprofit organizations, etc. Similar studies should be conducted in other countries as well so that we can better understand global patterns. We would expect the ideal conditions proposed by this theory to be better satisfied in nonprofit organizations such as universities, charitable foundations, and government agencies.

Further careful empirical and theoretical studies are needed to understand all these issues much better. We view our theory only as a start of a new direction of inquiry in this important area.

As noted, our theory, as it stands, is applicable only for large corporations, with tens of thousands of employees, and is not applicable to small companies, startups, and so on. It is not applicable to the entertainment industry where a few select individuals, as we discussed above, can create much of the value in their product. So this is another natural next step in the development of this theory.

## 7.2.4 How Much of the Success is due to an Amazing Economic Ecosystem?

The discussion above naturally leads us to the broader question. How do we value and compensate for the invaluable role played by the economic ecosystem that empowers entrepreneurs and companies to succeed immensely?

For example, consider all the great American entrepreneurial success stories in the computer and Internet era, from Shockley Semiconductor, Fairchild Semiconductor, and Intel to Apple, Microsoft, Google, and Facebook, and the thousands of other companies, big and small, over a span of about sixty years—essentially the story of Silicon Valley. How much of this success is due to the great ideas and the hard work of the entrepreneurs, and how much is due to the American capitalist society that made this possible?

Simply put, could these great companies, and the jaw-dropping personal fortunes, have been created by their hard-working innovative entrepreneurs anywhere else in the world? Could they have been founded in China, India, Africa, or in the Middle East, some fifty years ago when Intel was founded, or forty years ago when Apple and Microsoft were founded, or even twenty years back when Google was founded?

It is highly unlikely. It would not have been possible even in Europe, which is a lot closer to the U.S. ethos in many ways than the others. One can confidently say that even the cofounders of these great companies would agree it would not have been possible anywhere else, because for such ideas and efforts to succeed on that scale, one needs a highly productive economic ecosystem, created and nurtured by a democratic capitalist society, which provides a wide range of essential services that are absolutely necessary for such successes to occur. All this is not to take away from the great contributions of the pioneers, but we need to have an honest discussion about how and where such successes came about.

Of course, China, India, and other countries are becoming more hospitable to innovation and entrepreneurship these days, inspired in part by the success of Silicon Valley, and are giving birth to great companies like Alibaba, but this was not always the case in the last fifty years. Despite such recent progress, the United States in general, and Silicon Valley in particular, still remains the hotbed of entrepreneurship globally.

If we examined the characteristics of this extraordinarily enabling ecosystem that gave birth to Silicon Valley, we would readily recognize its vital elements. A partial list would comprise (not necessarily in any particular order of importance) a large pool of talented and driven workforce, nurtured and educated at great schools and universities; embedded in dynamic but stable communities; supported by a wide variety of infrastructure and services (to address food, shelter, water, energy, financial, and other needs), in a liberal, vibrant, risk-taking entrepreneurial culture of innovation and wealth creation; supported by a hospitable living environment maintained, protected, and served by a variety of law and order services (sanitation, police, etc.); in a democratic society that protects equality, liberty, and the other basic rights, including intellectual property rights, of all citizens under the rule of law. We recognize that this is essentially a description of the U.S. society at large. While perhaps they are not as great a success story as the Silicon Valley, we have other such entrepreneurship-friendly environments in the United States, notably in New York City and Boston.

Without such an ecosystem nurtured and supported by the society at large, Facebook, Apple, Google, or any of the other great companies could not have emerged, no matter how greatly talented and hardworking the founders were.

Every year senior government officials from China, India, Singapore, and various European countries visit Silicon Valley to understand and learn its secrets for this amazing success over several decades. They want to reproduce such an ecosystem in their own countries. But, as we all know, this is not that easy for it requires all the elements listed above to exist, more or less, for a such a phenomenon to take off. And this doesn't happen overnight.

So ideas and hard work in themselves don't automatically translate into huge commercial successes and immense fortunes. They are necessary elements, but not sufficient. All the other enabling elements have to exist to go from a great idea to a great company to a great fortune. Some elements are admittedly more important than others, but one needs all of them. Such an ecosystem is a living system similar to a human being—one needs all the components to make it complete and fully functional, stable and happy, even though some parts, like the brain and heart, are more important than others. We call such an outcome *systemic success*, a perfect storm of good things, as opposed to *systemic failure*, a perfect storm of bad things, which we evidenced in the Global Financial Crisis.

So, how do we recognize and reward this extraordinarily productive free-market ecosystem? When these wonderful companies and their talented entrepreneurs succeeded financially, this ecosystem, which empowered such a marvelous outcome, also succeeded. This ecosystem is a silent partner. So where is this partner's fair share when the immense fortunes are distributed among the founders, the venture capitalists, and the employees of such companies? Shouldn't some of this go back to the ecosystem, back to the society, which made all this possible? This is not only the fair thing to do, from a moral perspective, but also the smart thing to do, from a utilitarian perspective, because you need this ecosystem alive and well so that the entrepreneurial successes could continue into the future.

One could argue that the return to the society is in the form of the many benefits its citizens enjoy through the technological advances of these companies. For example, the breakthroughs in computer and communication technologies that resulted in powerful but inexpensive computers and smart phones, or the variety of medicines and medical procedures that are routinely available, have greatly improved the quality of lives of millions of people. Isn't this a return on the investment made by society?

Yes, but only partially, because this is not as straightforward as one might think. We should remember that the companies also *need* the citizens as their potential customers for they are the market for companies' products and services. This is a symbiotic relationship. Thus, people pay for the benefits when they purchase the products and services. It is important to recognize that society plays two different roles in this context—as a *customer* or *market* for the products and services as well as an *investor* by enabling the creation of the companies.

Currently, we don't have a quantitative theory, or even empirical models, which can assess the value of the investment contribution of a society to the successes of its corporations. We don't have a theory that can delineate the return to society as a customer versus investor. Our theory might be considered as a step in that direction. Such a theory would justify both morally and economically the need for, and the amount of, corporate and personal taxes as a fair compensation for the contributions made by a society in its role as an investor. This moral and economic justification is apparently lost on those people who proudly and loudly proclaim that "taxation is theft" and that they believe in paying as little tax as possible. In an era of corporate inversion and a narrowly defined mission of corporations focused on short-term gains, our message may fall on deaf ears, which makes it all the more important to keep hammering it. This patently obvious recognition is sorely missing in the current myopic "winner take all" culture of the U.S. version of capitalism.

President Teddy Roosevelt recognized this more than a century ago when he observed, as quoted at the beginning of chapter 6, "The man of great wealth owes a peculiar obligation to the state because he derives special advantages from the mere existence of government." Herb Simon was more direct, as quoted earlier, when he said, "If we are generous with ourselves, I suppose we might claim that we 'earned' as much as one-fifth of our income. The rest is the patrimony associated with being a member of an enormously productive social system." Simon might perhaps be wrong about the percentage, but he was certainly not wrong about the underlying logic.

Defending the winner-take-all practice, some argue that the extraordinary paydays are essential to motivate people to take risks and innovate. This seems dubious and self-serving as it underplays the fact that entrepreneurs are often driven as much by passion to create something novel, to create something of value and impact, as by the monetary rewards. Again, Silicon Valley is full of such stories. Intel, Apple, Microsoft, Google, Facebook, and others were started by pioneers driven by passion. They would have done what they did even if they had ended up one-tenth or even one-hundredth as wealthy. Innovators are also motivated by peer recognition, as Steve Wozniak, creator of the first Apple computer, recounts. Speaking to National Public Radio in 2006, he explained that

> When I built this Apple I, the first computer to say a computer should look like a typewriter—it should have a keyboard—and the output device is a TV set, it wasn't really to show the world [that] here is the direction [it] should go [in]. It was to really show the people around me, to boast, to be clever, to get acknowledgment for having designed a very inexpensive computer.... My idea was never to sell anything. It was really to give it out.

Even those who are after monetary rewards have only a vague idea that they could get rich, perhaps make millions; but they typically don't know how much they could make for the simple reason no one has any idea how successful a particular company is going to be. It's anybody's guess at the start-up stage. It is not as if Jobs, Wozniak, Gates, and others like them would have said, "What? I will only make a few million dollars and not billions? Then, forget about it, I am not going to start this company. I am going to go and work for somebody else." Thus, to say that if we don't have have a winner-take-all culture that rewards hundreds of millions or billions of dollars, people will simply stop taking risks and innovating rings hollow. The fact that while financial rewards are important, they are not often the primary motivation seen from the lives of many wealthy tech entrepreneurs who continue to work and innovate. Money is no longer the motivation, but creating something novel, something valuable, something that can change the world is. No amount of money can even come close to the exhilaration and satisfaction one feels from creating something unique and valuable that makes a huge contribution and has a positive impact on people's lives.

So the refrain that we need "winner-take-all" paydays of exorbitant amounts, because otherwise the free-market genie will not come out of the bottle to perform its magic, is greatly exaggerated. Obviously financial incentives are needed, but they don't have to be winner-take-all greedy outcomes. If anything, going down that path is a dangerous strategy. To invoke another analogy, this path of extreme inequality and unfairness would lead to killing the goose—free-market capitalism—that lays the golden egg of wealth creation through innovation. The current version of U.S. capitalism needs a serious makeover, along the lines of *Conscious Capitalism*, which was the original vision of Adam Smith. As Bill George stated, "It is the only authentic form of capitalism."

Next, in chapter 8, we briefly review the different strands of our theory, synthesize and express the unified framework as formal postulates, and clarify certain subtle issues. We discuss the significance of the parameters $\alpha$, $\beta$, and $\gamma$. We also discuss the limitations of our theory, particularly with respect to the workings of a real-life free market, and we outline future directions in both theoretical and empirical studies to develop the theory further.

# CHAPTER EIGHT

# Final Synthesis and Future Directions

> I want to know His thoughts, the rest are details.
> ALBERT EINSTEIN

> If I have seen further it is by standing on the shoulders of giants.
> ISAAC NEWTON

The objective of this book is (1) to describe a new quantitative, analytical theory of fairness in economics and political philosophy for free-market capitalist societies and (2) to make testable quantitative predictions that can be verified by empirical data. The dominant theories of fairness, proposed by pioneering political philosophers such as Mill, Rawls, Nozick, and Dworkin, are all qualitative. These theories, no doubt, greatly deepened our understanding of liberty, equality, and justice. However, because of their purely qualitative nature they can't make testable quantitative predictions that can be verified with empirical data from real-life societies.

Such predictions would be particularly useful in the area of distributive justice in dealing with inequalities in income and wealth. While there is extensive empirical literature on income and wealth distributions, what is needed is a *normative analytical theory* for determining the fair distribution of income (and wealth, but we don't address this aspect here), even if it is limited to only an ideal free-market society. Knowing this distribution can serve as a fundamental benchmark against which we can evaluate the distributions seen in real life. In the absence of such a reference framework, the conclusions we reach by relying on empirical observations alone are likely to be incomplete, in an important manner. This benchmark can help us measure, and better understand, the deviations caused by nonidealities in the real world and develop appropriate policy frameworks to try to correct the inequalities. It can give us a quantitative basis for understanding and developing rational tax and transfer policies, pay packages for executives, and so on. This is the challenge we address with our theory.

In our approach, we start with the empirical reality that none of the pure model societies—purely utilitarian (à la Mill), egalitarian (à la Rawls or Dworkin), or libertarian (à la Nozick)—exist in the real world. What we find instead are hybrid societies that have combined features from these different philosophies, to varying degrees, as a political compromise to satisfy various constituencies. They typically incorporate Nozick-like free-market principles, suitably modified by a Rawlsian difference principle—like tax and transfer policies while making utilitarian cost-benefit trade-offs. Some societies have more of the egalitarian elements incorporated in them, such as in the Scandinavian countries. Others have less, as in the United States. Still others are somewhere in between on this spectrum.

We need a quantitative theory that can deal with this reality. We have presented such a theory in this book. Our theory integrates fundamental concepts from game theory, statistical mechanics, information theory, and systems engineering with foundational principles of political economy, as an alternative to the purely qualitative political economic theories of Mill, Rawls, Nozick, and Dworkin. Our theory is a hybrid animal—part human, part beast—combining the humane egalitarian features with the "survival of the fittest" free-market doctrine. This human-beast, this *Narasimha Avatar*, may not be as elegant as the pure creations of Mill, Rawls, Nozick, and Dworkin, but it has the potential of being more relevant and useful for crafting economic and corporate governance policies.

## 8.1 Mathematical Foundations of a Utopian Capitalist Society: Maximum Fairness as the Design Principle

We consider an ideal version of such a hybrid society, a utopia called BhuVai, explore its distributive justice property from the income distribution perspective, and compare it with real-world data to assess how close reality is to ideality. This hybrid utopia serves as our reference, as the desired target society where most people would love to live, one that can also be modeled mathematically for its distributive justice properties. We don't offer any particular philosophical defense or justification for our hybrid society, along the lines of Rawls, Nozick, or Dworkin, other than to defend it as a pragmatic solution commonly arrived at by citizens in real life to challenging trade-offs among liberty, fairness, and utility. That said, we do offer a *system-theoretic* rationale, founded on the principle of maximum entropy, as we discuss below. One might call our hybrid theory *pragmatic egalitarianism* or *pragmatic libertarianism*. One might also view this as a systems engineering perspective of a socioeconomic system, which adopts the engineer's approach to solve an important practical problem, utilizing whatever conceptual and modeling tools are available without worrying too much about philosophical niceties and economic orthodoxy.

In our framework, we let system-theoretic principles guide us in "designing" our model hybrid society. They help us identify the relevant components as well as the appropriate design and operational principles of this society, such as the intrinsic and extrinsic properties of our rational agents, microstates and macrostates of the society, phase space and accessible regions, evolution of the societal dynamics, equilibrium and stability, and equality, fairness, and morality in the society. Like Rawls, we also founded our hybrid society on two fundamental principles, liberty and equality. Thus, all people in our hybrid utopia are equally endowed with inalienable basic liberties. This is similar to assigning the same intrinsic properties to all agents in statistical mechanics and statistical teleodynamics (see chapters 4 and 5).

Regarding Rawls's second principle, first we focus on the difference principle part of it. In this, Rawls stipulates that a just society is one wherein the social and economic inequalities are to be arranged such that they are to be of the greatest benefit to the least-advantaged members of society. He proposes the maximin strategy to accomplish this. As Nozick argued, this treats the weaker members preferentially over the rest, which is unfair, thereby violating the equality principle. In our theory, we let the free market distribute the income, as Nozick recommends, and arrive at whatever income inequality that results naturally. Nozick does not mention what this distribution is in his work.

In our theory, however, we prove that the free market accomplishes this by maximizing fairness, under ideal conditions, yielding the lognormal distribution of income at equilibrium. Since we are maximizing fairness, we are not favoring one group over another; all are treated fairly, as opposed to Rawls's maximin strategy. For Nozick, the free-market outcome is fair because it was the result of legitimate transactions, jointly and voluntarily agreed to, among consenting individuals. He considered it fair because it was arrived at freely, without any coercion, by its participants.

In our framework, we prove, using both information theory and statistical mechanics, that this is *indeed the fairest outcome that is theoretically possible*, under ideal conditions. In that sense, Nozick was more right than he perhaps realized! He trusted the ideal free market to do the right thing, as Adam Smith did (1976). We have proved that his trust, and Smith's, were not misplaced in an ideal free market. Unfortunately, the real-world free market is a different story—we practice what Joseph Stiglitz refers to as *ersatz capitalism*. This crony capitalism of real-life free markets violates a number of our ideality assumptions regarding rent seeking, market and managerial power, labor's influence, and so on.

We also add a Rawlsian twist to this Nozickian free market–driven distribution. We introduce a minimum wage as a floor to this distribution (as our calculations show in chapters 6 and 7) to protect the working poor. This compromise solution reflects the reality practiced in most countries, including the United States, which have a minimum-wage constraint.

Finally, regarding Rawls's *fair equality of opportunity principle*—all positions and offices are open to all—is equivalent to the fundamental statistical teleodynamic principle that all parts of the accessible phase space are equally available to all qualified agents in our theory. This is the direct consequence of the fundamental postulate in statistical mechanics, the *equal a priori probability postulate*, which is again a fairness requirement.

Thus, both fundamental principles of Rawls are present in our theory (except for the maximin algorithm). They arise naturally in the statistical teleodynamics framework, as part of its systems design protocol. We did not incorporate them because Rawls said so. The principle of maximum fairness, i.e., the principle of maximum entropy, plays a central role in making these assignments. Whether it is the set of intrinsic properties of inalienable rights and liberties, or the fair equality of opportunities, or the fairest economic inequality, they all are *determined by maximum fairness as the design principle*.[1]

Furthermore, not only do we retain all the key features of Rawls we have also managed to incorporate Nozick as well in an internally consistent matter. Even though we are mixing and matching two different philosophies, Rawls's egalitarian tradition with Nozick's libertarian one, there is no inconsistency here (at least in the context of distributive justice) because they all are arrived at using the same principle of maximum fairness, the principle of maximum entropy. Thus, contrary to what typically happens when you mix two opposing traditions, our hybrid society is internally consistent in its foundational principles, which is indeed reassuring. Our Narasimha Avatar is beautiful, after all, not an ugly mishmash of conflicting components! This is the system-theoretic justification we mentioned above, which goes beyond the defense based on pragmatism alone.[2]

Another vital feature in our theory is the societal objective of equality of welfare for all its citizens in our hybrid utopia, BhuVai. Once again, this is also grounded on our principle of maximum fairness. In BhuVai, everyone is equally happy—no one is happier than another. What else could be fairer? As a result, there is no envy. No one would want to trade her life for someone else's, for everyone is living her life the way she is inclined to, enjoying the same level of happiness as everyone else.

In summary, our utopia is founded on one simple overarching design principle—the principle of equality, which we call as the principle of maximum fairness. This manifests itself in three fundamental ways: (1) equality of basic liberties for everyone, (2) equality of opportunity and engagement in a competitive free market, and (3) equality of welfare or happiness.

HOW USEFUL IS THIS THEORY IN THE REAL WORLD? One might think that this is all wonderful, but it is completely unrealistic and therefore totally useless. First, no one could measure, in practice, happiness correctly and consistently across a population, not to mention how such a policy might be implemented. This would be one's typical gut reaction.

But as we showed in chapter 6, this unreachable ideal, this unreachable target, is surprisingly useful in engineering real-world societies. Since this is the utopia everyone would love to live in, were a real-world society to come close to this utopia, it would be doing really well for its citizens.

As we all know well, we do not expect real-world societies to be this utopian one-class society, i.e., a society with no class structure. A janitor and a CEO do not enjoy the same effective utility in the real world—this could happen only in a utopian one-class society. Real societies are likely to be many-class societies, with perhaps every individual having different values of $\alpha$, $\beta$, and $\gamma$ (see also the discussion in section 8.4). But as we discussed in section 6.1, they may be usefully grouped as three-class or four-class societies, such as in lower, middle, upper, and wealthy classes. The welfare enjoyed by each class is the same for everyone in that class. However, it increases progressively as we move up the class hierarchy. This is signaled by the emergence of *four different, nonoverlapping, lognormal distributions*, as we have shown. So a janitor and a CEO do enjoy substantially different utilities, in reality, as they belong to different lognormal distributions.

From a fairness perspective, this is not a desirable outcome, as we have described. If, however, a society manages to have its overall income distribution as a combination of *two* nonoverlapping lognormal distributions, with the lower lognormal representing an overwhelming majority of its population, with its utility close to that of the upper lognormal, then it is approaching utopian-level fairness. This is what the Scandinavian countries seem to have accomplished based on *aggregate income share* data (see figures 6.3 to 6.5; more thorough statistical analysis is needed to understand this better).

Thus, the one-class utopia model, while unrealistic in practice, surprisingly serves a very useful purpose by defining the desired target, the ideal benchmark, the gold standard, in quantitative terms. This quantitative metric can guide us to design better tax and transfer policies, executive compensation packages, etc., to correct the income inequality in a society by getting as close to ideality as desired by its citizens in practice, given the society's economic and political realities.

In BhuVai something interesting happens. All philosophies are satisfied by the same outcome, as far as distributive justice is concerned. We achieve the utilitarian dream of greater good, i.e., maximizing societal utility or happiness, pursued by Bentham and Mill. Rawlsians should be satisfied, for even the weak are just as happy as the strong. Nozickians should be happy because we achieve this outcome via free-market libertarianism, not because of a heavy-handed state. Dworkinians should be happy because we achieved an envy-free society. So all paths converge on this destination. A true utopia indeed!

## Final Synthesis and Future Directions

KEY FEATURES OF STATISTICAL TELEODYNAMICS We summarize the key features of statistical teleodynamics and BhuVai:

- Statistical teleodynamics is founded on the fundamental principles of liberty and equality (as in Rawls, in Nozick, and in Dworkin).
- All individuals enjoy the same basic liberties (as in Rawls and in Nozick).
- There is fair equality of opportunity (as in Rawls).
- Distributive justice and economic inequalities are determined by the free market (as in Nozick), except for the minimum-wage constraint. Rawls justifies his difference principle by focusing on the *inputs* (i.e., natural endowments of individuals) to strive toward equality by correcting the inequities in the initial assignment. Our theory does not require this as it focuses on the *outcome* (i.e., economic welfare) and proves mathematically all are treated equally as they all achieve the same level of economic welfare despite differences in their natural endowments.
- Equality of welfare is the societal objective (not present in the work of Rawls, Nozick, or Dworkin, but generally desired by utilitarians).
- The moral justification is proved *analytically* for the free-market economy (not found in the theories of Smith, Mill, Rawls, Nozick, Dworkin, or others).
- Quantitative models integrate key concepts from game theory, statistical mechanics, information theory, and systems engineering with those of political economy. The central design and governing principle is the principle of maximum fairness, implemented as the principle of maximum entropy (the quantitative feature is absent in the philosophical theories).
- Stability and social optimality of the equilibrium state are proved (absent in the other theories).
- The importance of both design criteria—efficiency and *robustness*—for a free-market system is recognized and accounted for in accordance with systems engineering principles of design and operation, which is absent in the other theories. (In particular, mainstream economics emphasizes only Pareto efficiency while ignoring the robustness property altogether.)
- Testable predictions can be verified by empirical data (absent in the theories of Smith, Mill, Rawls, Nozick, and Dworkin).

- Income distribution at equilibrium is proved to be the lognormal distribution. This is validated by empirical data that suggest that the bottom ~95%–97% of the population follows this pattern.
- The ideal share of income is predicted for the bottom 90%, top 10%–1%, and top 1% of the population (or for any other percentile breakdown one desires).
- Comparisons of the income share predictions with real-world data from twelve countries reveal interesting results, in particular about how the Scandinavian countries operate as near-ideal free markets for the bottom 99% of their population.

### 8.1.1 Maximum Fairness as a Symmetry Principle

The common motif that connects all these disparate fields—game theory, statistical mechanics, information theory, systems engineering, economics, and political philosophy—is the principle of maximum fairness, modeled by the true meaning of entropy. Fairness is what Bernoulli and Laplace were modeling in their principle of insufficient reason—if the $n$ possible outcomes are indistinguishable except for their names, then each outcome should be assigned a probability equal to $1/n$. This is the fairest treatment of all outcomes, by not preferring one outcome over another without a reason. As noted, the generalized version of this principle of maximum fairness is the principle of maximum entropy.

It is interesting, at this juncture, to note a parallel between the role of equality and fairness in the socioeconomic sphere and in the material universe.

The material universe, too, is founded on the principle of equality. This is reflected in the different symmetries that are the basis of its laws. A symmetry, simply put, refers to some regularity or "sameness" in an object. It is the property of remaining invariant, i.e., changeless, under a transformation such as translation or orientation in space. For example, a circle remains the same when you rotate it about its center. No one could tell the difference between the original circle and the rotated circle because they look exactly the same. This property is called the *rotational symmetry* of circle.

Similarly, it turns out that that if you place a system of particles in empty space, far away from anything that might affect it, it does not matter where exactly you put it. That is, there are no preferred locations in empty space—*all locations are equivalent*. This symmetry is called the *translational symmetry* of empty space, and this leads to the law of conservation of linear momentum. Similarly, if you place a system of particles in empty space, it does not make a difference at what angle you place it. There are no preferred directions or orientations in empty space—*all directions are equivalent*. This *rotational symmetry* leads to the law of conservation of angular momentum. Similarly, *translational symmetry in time* leads to the law of conservation of energy. A property called *gauge symmetry* leads to the law of conservation of electric charge.

This connection between symmetries and conservation laws is one of the most beautiful and profound discoveries in all of physics, and it lies at the very foundation of the modern field-theoretic models of the universe. This pioneering discovery, known as *Noether's theorem*, is due to Emmy Noether, an eminent German mathematician admired by many, including Hilbert and Einstein, who made the discovery in 1915 to 1918 (Byers 1998). Leon M. Lederman, Nobel Laureate in Physics, and Christopher T. Hill write in their book *Symmetry and the Beautiful Universe* (Lederman and Hill 2004, 23–24): "Emmy Noether made one of the most significant contributions to human knowledge through her remarkable theorem.... One might argue that Noether's theorem is as important to understanding the dynamical laws of nature as is the Pythagorean theorem to geometry."

By requiring the equality of all inertial frames of reference as the foundational principle, Einstein derived his special theory of relativity. By generalizing this equality principle to include noninertial frames of reference, he went on further to develop the general theory of relativity and the theory of gravitation.

Interestingly, in statistical mechanics, the equality of all accessible phase space points, a kind of translational symmetry in phase space, does not lead to a conservation law but to a maximum principle, namely, maximum entropy.

Thus, we see that behind the fundamental laws of physics, there are symmetry principles, and behind the symmetry principles, there is the principle of equality.[3] Thus, the design principle of the universe, i.e., the design of its fundamental laws, is founded on the principle of equality.

Treating everything equally is nothing other than exercising maximum fairness. No location in space is to be preferred over another. No orientation in space is to be preferred over another. And so on. Not preferring one thing over another for no reason is the very definition of fairness. *Thus, at its most fundamental level, the universe is founded on the principle of maximum fairness.* Maximum fairness is quite literally a *universal principle.*

Equality as a design principle puts an end to the series of why questions we could ask. Why should energy be conserved? Because of translational symmetry in time; i.e., all points in time are equivalent. Why should they be equivalent? Because there is no reason for them not to be, i.e., out of fairness. Now there is no more why left to pursue.

It is therefore quite comforting to have the principle of equality, the principle of maximum fairness, as the design principle of our hybrid utopian society BhuVai as well.

In systems engineering, one recognizes, as we discussed in chapter 2, that the purpose of a system is to take some input and transform it into desirable output as optimally and safely as possible. In that sense, we note that our principle of equality is enforced, from a *system-theoretic perspective*, in three fundamental ways in the context of distributive justice: (1) *Equality in input* is the equality of all citizens in their endowments of their basic liberties. (2) *Equality in opportunity to compete* means that every citizen has the opportunity to compete in an ideal free-market environment to market her products and services based on her merits and contributions and is not limited by any kind of discrimination other than pure merit. (3) *Equality in outcome*, as a result of items 1 and 2, means that all citizens enjoy the same welfare, i.e., the equality of welfare. Thus, the principle of equality is central to the three fundamental components of free-market dynamics in our system-theoretic perspective of utopia: *input, transformation*, and *output.* Our mathematical framework demonstrates how these three equalities are executed in an ideal free-market environment.

While it is particularly appealing to find that our ideal society is also organized by the principle of maximum fairness, much as the material universe is, there is one important difference. Molecules always obey the underlying laws and principles, humans don't. Molecules don't hijack the system, people do.

Therefore, we need to take extra efforts to ensure maximum fairness in a free-market society. It won't necessarily happen by itself all the time. Recall Reich's (2015, 3) caution: "Few ideas have more profoundly poisoned the minds of more people than the notion of a 'free market' existing somewhere in the universe, into which government 'intrudes.'... A market—any market—requires that government make and enforce the rules of the game. In most modern democracies, such rules emanate from legislatures, administrative agencies, and courts. Government doesn't 'intrude' on the 'free market.' It creates the market." Hence, we need appropriate rules and regulations to enforce fairness and to ensure that the resulting free market behaves ideally.

## 8.2  Statistical Teleodynamics: Macroview

We now provide a formal statement of our theory in two related, but different, perspectives: the *top-down* and *bottom-up* perspectives. One may think of these as the macroview and microview of the system, respectively. The macroview focuses on the "design" specifications of the overall system (i.e., a society), its goal(s), and the associated constraints imposed by the environment. This is the information-theoretic perspective of the system.

A more extensive discussion of statistical teleodynamics and its relation to different design aspects of complex teleological systems, such as network topology, can be found in my earlier essay (Venkatasubramanian 2007). Since including all that here might take us far from the main theme of the book, namely, modeling income inequality, we refrain from doing so.

We first consider the macroview, expressed as four postulates. These would correspond to similar postulates in thermodynamics; in fact, the former reduce to the latter for the case of purpose-free entities such as molecules, as we discuss below.

Building on our earlier work (Venkatasubramanian 2007, 2010), we list the fundamental postulates of statistical teleodynamics:

1. *Macro survival postulate.* A teleological system is designed, or it evolves, to meet its survival objective(s) $\Xi(\xi_1, \xi_2, \xi_3, \ldots)$ in its operating environment, where $\Xi$ is the overall *fitness* for survival, which is a function of a number of survival-related parameters $\xi_1$, $\xi_2$, $\xi_3$, etc.

   For example, as noted, the objective of a corporation is to make money for its investors and shareholders. If it fails to do that, it won't survive. In the context of our discussion about purely utilitarian, egalitarian, and libertarian societies (chapter 2), this postulate formally states why we do not see these societies in real life, but only their hybrid avatars. The hybrid avatars are better fit to meet the constraints of the environment, such as the conflicting demands from their electorate.

2. *Performance target postulate.* The operating environment imposes both qualitative and quantitative performance constraint(s) $\Theta(\theta_1, \theta_2, \theta_3, \ldots)$ that the teleological system must satisfy for its survival.

   To survive, the teleological system must execute various functions at some targeted levels of performance or fitness $\Theta$, which are the design constraint(s) (e.g., $\langle \Xi \rangle = \Theta$) demanded by the environment.

   For instance, if a corporation delivers a ROI of only 5% in a market environment where the average is 15%, it won't survive. It will likely be acquired, or taken over and stripped of its assets and sold, and so on. In this example, $\Xi$ is the ROI and $\Theta = 15\%$. Thus, the environment determines *what matters* (e.g., ROI) and *how much* (e.g., 15%) for a system's survival. That is, the environment imposes both qualitative (i.e., what matters) and quantitative (i.e., how much) constraints. These might be different for different market segments.

   In essence, this postulate states that the environment demands *Pareto optimality* from the system for its survival, emphasizing the need for an efficient performance.

   Similarly, for hybrid societies, the local culture and traditions, conditioned by decades, if not centuries, of experience, determine what matters and how much in choosing the trade-offs among utilitarianism, egalitarianism, and libertarianism. So in Europe (particularly in Scandinavia) the parameters are set empirically through trial-and-error adaptation, in tilting the society more toward egalitarianism while in the United States it is biased more toward libertarianism.

3. *Optimally robust design postulate.* A teleological system is designed, or it evolves, to be optimally robust in order to meet its survival objective(s) in all potential operating environments.

   Since the nature of future operating environments in which a teleological system must survive is typically unknown, often unknowable, and hence *uncertain*, the system must be designed (or evolved) in such a way that it will be able to execute all its functions, achieve its performance targets, and meet its survival objective(s) in the widest set of possible operating environments.

In other words, a company, for instance, should be able to weather a wide variety of market conditions, survive, and grow. That is, the company should be stable and functional even under disruptions. Similarly, one would prefer to live in a society that has demonstrated its ability to survive and thrive, over a long time, despite all kinds of economic, social, or political turmoil. Such a society is *optimally robust* or *optimally stable*. This is essentially the same as what we discussed in sections 5.5.2, 5.9, and 5.10.

This postulate ensures that the system is not fragile while meeting the demands of postulates 1 and 2. The first two do not ensure survival under a wide variety of conditions. You might consider the first two postulates *minimum requirements*, necessary conditions, imposed on the system. But to survive over a long time, under a variety of threats, it needs to satisfy the third postulate as well.

As a simple example, consider an express delivery service company that operates in the southeastern United States. It owns a number of planes and has its own flight network. To achieve speed of delivery, it needs to fly the packages from one city to another either nonstop or with at most one stopover and a plane change. To meet this requirement while keeping costs down, the airline creates one hub at a major city in the southeast. This guarantees that a package from any city X to any other city Y in its operational area can be delivered with at most one stopover. If there is a direct flight between X and Y, it is done nonstop. Otherwise, the package will fly from X to the central hub and then from there to Y. This configuration is known as the *star network* as the hub-and-spokes flights of the network look like a star. Let us say that, in general, 20% of the packages are delivered with no stops and the rest with one stopover. This would satisfy the performance demanded by the environment (i.e., the market) in postulates 1 and 2.

Now consider the situation where a severe weather event such as a tornado or a hurricane damages the hub seriously. The other nodes (known as the *leaf* nodes) of the network are not affected. They can collect the packages, and the planes can take off from there and deliver them wherever they have direct flights. But for the other cities, they have to go through the hub, which they can't do now. So about 80% of the market will not be served for several weeks, until the hub is repaired and ready. This could cause serious losses to the company, and it might even cause the company to file for bankruptcy.

Thus, although the company was able to meet the market's metrics, under normal conditions, its network design was not robust to disruptions. In other words, the company was *Pareto-efficient* only under normal conditions. A better design would have been to have two hubs, not just one, in two different cities to minimize risk. Three would have been even better. This, of course, would increase the initial investment cost, but then the company can potentially avoid large operational losses due to major disruptions. (As an aside, FedEx has one national hub in Indianapolis and five other regional hubs spread around the country.)

Thus, system engineers would consider such possibilities and build in some redundancy in the network, to the extent allowed by cost constraints, to minimize risk. This is what we mean by robust design. This is similar to the automobile design example we discussed in section 5.9.

Therefore, the third postulate is also necessary, in addition to the first two, because of such uncertainties in the environment.

4. *Maximum entropy design postulate.* The overall design that maximizes entropy subject to the design criteria in postulate 2 is the optimally robust design of the teleological system, where the optimality is with respect to all potential operating environments.

The maximum entropy design is *optimal* because this design accommodates *maximum uncertainty* about the future operating environments. Note that for a particular operating environment, there might be some other network (e.g., the star network) that could outperform the maximum entropy network with respect to the performance target. However, such a biased design might fail when the operating environment is changed (e.g., disruption due to a weather event), whereas the maximum entropy-based network is more likely to survive and continue to perform. The "star" network design is biased in the sense that it assumed normal operation at all times, i.e., it is biased towards normalcy and did not anticipate disruptions.

However, under entropy maximization, the system's performance is optimized to accommodate a wide variety of future operating conditions whose nature is unknown, unknowable, and hence uncertain. Therefore, one designs by allowing for maximum uncertainty, which is what maximum entropy does. This is what we mean by an optimally robust design. Such a design makes the optimal trade-off between Pareto efficiency and robustness under the given constraints imposed by the environment.

In this engineering example, we have utilized the interpretation of entropy as a measure of uncertainty in reasoning about the stability of the system. When the system is a corporation or a free-market society, with utility-maximizing fairness-seeking agents, its stability is determined by (as we discussed above and in chapters 2, 3, and 5) the fairness in the system.

Consider the two extreme compensation scenarios we examined in chapter 5. In the equality of pay scenario, the company became *unstable* as employees started leaving because many felt they were being compensated *unfairly*. Stability (i.e., survival) could be restored only by changing its pay distribution and making it fair for everyone. The same is also true for the other extreme scenario, when the CEO got most of the pay and the rest got a pittance. Again, the company became unstable due to the unfairness in the pay distribution.

In an ideal free-market democratic system, where people are free to participate or not, people would not continue to participate if they felt they were not getting a fair deal, and the system would collapse. They might tolerate the unfairness up to a point, but beyond some threshold they would refuse to cooperate and the system would collapse.

We recall that this is the point both Scanlon and Reich make (see section 1.4). Recall Scanlon's remarks: "These are not just objections to inequality and its consequences: they are at the same time challenges to the legitimacy of the system itself. The holdings of the rich are not legitimate if they are acquired through competition from which others are excluded, and made possible by laws that are shaped by the rich for the benefit of the rich. In these ways, economic inequality can undermine the conditions of its own legitimacy." Similarly, Reich (2015, 167) cautioned: "In summary, when people feel that the system is unfair and arbitrary and that hard work does not pay off, we all end up losing.... When capitalism ceases to deliver economic gains to the majority, it eventually stops delivering them at all—even to a wealthy minority at the top."

In an authoritarian society it is a different story. There stability is maintained by external force. But for a society to be stable *on its own*, in a self-organized manner, and perform despite disruptions, its participants need to feel they are getting a fair deal, at least most of the time. That's only when they would play ball and cooperate. Therefore, societal stability is determined by fairness, and the criterion for maximizing stability is maximizing fairness, which is the same as maximizing entropy. Hence, this postulate may also be called *maximum fairness in design postulate*.

As we proved in section 5.9, the equilibrium state we arrive at by maximizing the game-theoretic potential $\phi$, i.e., by maximizing entropy, is an *asymptotically stable equilibrium*. This mathematical criterion confirms our intuitive expectation that for a society to be maximally stable, it needs to maximize fairness; i.e., every citizen feels he has received a fair deal.

In section 5.5.3 what we discussed is essentially the application of these four postulates to answer this question: What is the least biased distribution of $M$ dollars among $N$ employees given the constraints $E[\ln S] = \mu$ and $E[(\ln S)^2] = \sigma^2$? The constraints are specified by the first and second postulates, and the least biased distribution is the application of the fourth postulate. The third postulate is implied as the fourth postulate is the solution to the third.

For nonteleological, purpose-free, entities, such as the ones seen in thermodynamics (e.g., gas molecules in a pressure vessel), the first postulate is not relevant as there is no survival objective targeted by the entities. Similarly, the third postulate is also not applicable. The second postulate regarding the constraints imposed by the environment simply reduces to the conservation of energy requirement, which, of course, is the first law of thermodynamics. The fourth postulate also simplifies to the conventional maximum entropy requirement, which is the second law of thermodynamics. Thus, we recover the conventional thermodynamic laws as a limiting case of statistical teleodynamics for purpose-free thermodynamic systems. Again, this internal consistency is reassuring.

STATISTICAL TELEODYNAMICS AND DARWINIAN EVOLUTION As one can readily see, there is considerable overlap between our theory and Darwin's (1859) theory of evolution. In fact, postulates 1 and 2 are very much Darwinian. However, in the traditional Darwinian framework there is no precise articulation of the concept of robust design in *quantitative* terms. Furthermore, any connection to statistical mechanics or entropy is also missing. Darwinian theory recognizes, however, in qualitative terms, that the environment sets the fitness target, which is expressed through its survival of the fittest postulate.

What is new in our theory is that, through the third and fourth postulates, we precisely articulate the *robust design* concept *quantitatively*, in terms of *maximizing uncertainty* about the potential operating environments. In addition, we make the connection to statistical mechanics through the maximum entropy principle, thereby integrating the evolutionary biological universe, human-engineered teleological systems, and thermodynamical systems, all within the same unified conceptual framework. This unified framework makes us realize clearly the usefulness of the important result that the *maximum entropy design* is the *optimal design*, optimal for surviving and performing in all potential operating environments.

What we have proposed is a unification of the Darwinian perspective with Boltzmann-Gibbs's statistical thermodynamics framework, Shannon-Jaynes's information theory, and Nash-Shapley's potential game theory. In the Darwinian formulation, the entropic perspective grounded in statistical mechanics is absent. As a result, the Darwinian theory has come to be seen as a framework that is applicable only to evolutionary *living* systems. On the other hand, Boltzmann-Gibbs's statistical formulation of entropy is seen as a theory that is applicable only to *inanimate* thermodynamical systems. The traditional view has been that Boltzmann's perspective has little to say about living systems. In our work, however, we have shown how these two seemingly different perspectives may be reconciled into a unified framework with the maximum entropy principle playing a crucial role in linking the two.

In our framework, the Darwinian adaptive evolution of new species through genetic mutation and crossover operations is nothing other than nature (i.e., living systems) exploring, stochastically, the design space of potential candidates to create a diverse population of progeny, some of whom might survive if the current environment were to change and become more challenging. Since nature doesn't know (in fact, cannot know) in advance what the change might be like, its best strategy would be to design for maximum uncertainty under the current constraints. That is, implement the optimally robust design policy. This is what is implemented through its genetic exploration.

One may detect an interesting parallel here. First Shannon, and later Jaynes, generalized the concept of entropy from its restricted thermodynamic sense to the information-theoretic interpretation of the second law of thermodynamics. Likewise, our theory generalizes the conservation of energy $E$ constraint, i.e., $\langle E \rangle =$ constant, to the survival-related constraint of $\langle \Xi \rangle = \Theta$ for teleological systems, thereby expanding the scope of the essential principle behind the first law of thermodynamics.

## 8.3 Statistical Teleodynamics: Microview

As opposed to the macroview, which is the design perspective, the microview emphasizes the operations perspective of the system, focusing on the individual agents, their goals, interactions, and dynamical evolution. This is the *game theoretic-statistical mechanics* perspective we developed in chapters 3 and 5.

These are the two fundamental postulates of the microview:

1. *Micro survival postulate.* The goal of an individual agent is to survive and grow in a competitive environment, populated by other such agents, by continually taking actions to maximize its effective utility.

   We are using *survival* and *growth* in a broad sense here. They apply not only to a specific agent but also to its progeny. That is, an agent is interested in not only its own immediate survival and growth but also the long-term survival and growth of its descendants. So it maximizes its utility, keeping all these objectives in mind. Further, we also include the goal of pursuit of happiness as an integral part of the survival objective. For instance, if one is terribly unhappy and depressed, he may not value survival at all, as he may see no point in continuing to live. Hence, our use of the term *survival* stands for the broader set of objectives that an individual has. Of course, the happiness feature is irrelevant (as far as we know) regarding lower forms of life (e.g., cells, bacteria, viruses, fruit flies). We do not address the question of whether biological evolution is truly teleological. We are only interested in the narrow scope of the definition of the term *teleological* to mean that the lower forms of life behave as if they are driven by a survival objective. The higher forms, particularly intelligent life, are overtly teleological.

2. *Equilibrium postulate.* There are a number of ways of stating this postulate, and we list two versions that are appropriate for the problem at hand:

- A teleodynamic system of many competing rational agents reaches equilibrium when fairness in effective utility distribution is maximized.

  or

- A teleodynamic system of many competing rational agents reaches equilibrium when every agent enjoys the same level of effective utility.

The first postulate is simply a statement of the well-known *Homo Economicus* model commonly used in economic theories. It states what motivates the motion of agents in the phase space. The rational agents switch from one job to another, driven by their desire to maximize utility, i.e., exhibit *teleodynamic* (i.e., *goal-driven*) behavior.

The second postulate states that when the movements come to a "stop," that is a dynamic statistical equilibrium. As we discussed in sections 3.5, 5.4, and 5.5, at equilibrium the potential $\phi$, or fairness, is maximized. Thus, this postulate is the equivalent of the fundamental postulate in statistical mechanics, the *equal a priori probability postulate*. In statistical mechanics, we don't have the equivalent of the first postulate because it is assumed that molecules are constantly in motion driven by thermal agitation.

Thus, both the macroview and the microview of statistical teleodynamics describe the same phenomenon of statistical equilibrium from two different, but related, perspectives, arriving at the same conclusions, both mathematically and intuitively. As we have noted several times, the connection between the two views occurs through the principle of maximizing fairness modeled by the game-theoretic potential and entropy.

It is worth reiterating at this juncture the importance of the connection between statistical equilibrium and Nash equilibrium, particularly in the context of games with *mixed strategies* at Nash equilibrium. As is well known (Easley and Kleinberg 2010), in mixed-strategy games such as matching pennies and rock-paper-scissors, Nash equilibrium is reached when players mix up their strategies with *equal probabilities*. For example, it is $\frac{1}{2} : \frac{1}{2}$ in playing head or tail in the matching pennies game, $\frac{1}{3} : \frac{1}{3} : \frac{1}{3}$ in the rock-paper-scissors game. The point is that the strategy of playing different choices with equal probabilities will make the opponent *indifferent* to them, thereby removing his ability to exploit any particular choice. That is, each player wants his behavior to be maximally unpredictable to the other, so that his behavior can't be taken advantage of.

This is nothing but the *maximum entropy* strategy. When the uncertainty about the outcome (whether it is head or tail) is maximum (i.e., $\frac{1}{2} : \frac{1}{2}$ probabilities), entropy is maximized, per equation (4.2). Thus, Nash equilibrium is reached when entropy is maximized, which, as we recall, is the same requirement for reaching statistical equilibrium. Note the reference to *indifference* above, a term that game theorists use to explain and justify the mixed-strategy Nash equilibrium. Their use of this term and the associated justification are essentially the same as what is behind the principle of insufficient reason, i.e., the principle of indifference, discussed in section 5.6. Therefore, at Nash equilibrium players are maximizing the uncertainty about their moves (i.e., their unpredictability) or maximizing the robustness of the outcome to the potential downside of lower payoffs. As we discussed in section 5.6, both *uncertainty* and *robustness* are different interpretations of entropy.

Thus, once again, we see the deep connection between game theory and statistical mechanics, and between Nash equilibrium and statistical equilibrium, through the concept of maximum entropy. As the number agents ($N$) becomes very large, no agent can possibly know all the strategies of the other agents, anticipate their best strategies to be played, and then play an appropriately customized strategy. As a result, every agent plays its generic best strategy, which in this case is to acquire the best education, the best work experience, the best network of contacts, and so on, in order to compete effectively with other agents who have prepared themselves equally well. Thus, any given agent's prospect of winning in a population of $N$ peers is just $1/N$ as we saw in section 3.2. Thus, for large $N$, this game of strategy becomes essentially equivalent to a game of chance. Therefore, both game theory and statistical mechanics converge and lead to identical results.

STATISTICAL TELEODYNAMICS: A TRANSDISCIPLINARY SYNTHESIS As readers would no doubt recognize, several elements of the proposed theory are not new and have been around for a long time. However, we believe that this is the first time one has articulated the deep connections between the fundamental principles of political philosophy, free-market economics, statistical mechanics, information theory, game theory, and systems engineering through a coherent and unified theoretical framework. This unified framework is a synthesis of seemingly disparate fundamental principles from different domains, as summarized now:

- We follow Rawls in endowing our teleological agents with all the *basic liberties* and by treating all agents *equally*. (Fundamental principles from political philosophy)
- We follow Smith and Nozick in allowing the economic evolution of the agents to be determined through the self-organizing, adaptive, *free-market* dynamics. (Fundamental principle from political philosophy)
- We follow Smith, Mill, and Darwin in assigning a survival instinct to our agents which manifests as the behavior to maximize one's utility. (Fundamental principle from biology and political philosophy)
- We follow Johnson, Locke, Bentham, and Mill in defining our society's goal as happiness, but extend the objective as universal happiness; i.e., all agents enjoy the same level of happiness. (Fundamental principle from political philosophy)
- We identify entropy as a measure of fairness and demonstrate its critical role in free-market dynamics by building on the work of Clausius, Boltzmann, Gibbs, Shannon, and Jaynes. This paves the way to prove maximizing entropy as the condition for achieving statistical equilibrium in teleodynamics. (Fundamental principle from statistical mechanics and information theory)
- We also identify the game-theoretic *potential* as entropy, and hence a measure of fairness, by building on the work of Nash and Shapley. This paves the way for proving that statistical equilibrium is the same as Nash equilibrium. (Fundamental principle from game theory)

- We follow Lyapunov and prove the asymptotic stability of the free-market equilibrium. (Fundamental principle from dynamical systems theory)
- We recognize, and account for, the importance of both efficiency and robustness as the design and operational criteria for the ideal free market, and we identify that the robustness property is due to the maximization of fairness in the system (Fundamental principle from systems engineering).

## 8.4 Significance of the Utility Function and Its Parameters

We start with the premise that, in general, a society will have people with varying talents, skills, and desires. The goal of our ideal society BhuVai is to provide everyone with the necessary opportunities to acquire additional skills through education and experience so that they all can rise to their full potential, as allowed by their intrinsic abilities, engage in meaningful and challenging occupations, productively contribute at their maximum capacity to the society, and enjoy the same level of happiness as any other citizen in that society.

The one-class system models the utopian hybrid society BhuVai. It has a homogeneous population of people, who have the same utility or payoff preferences; i.e., the parameters $\alpha$, $\beta$, and $\gamma$ are the same for every agent and are constants independent of time or job category. Homogeneous *does not mean* that the people are identical in their education, talent, skills, abilities, experience, desires, and ambitions. It only means they all weight the utility from salary, disutility from employment, and utility from future prospects to the same extent. This society recognizes the dignity of labor, that everyone is important, making a valuable contribution to society, and deserving the same level of happiness as everyone else. That's why the payoff (i.e., happiness) is the same for everyone at equilibrium. The income inequality in this system at equilibrium is given by the lognormal distribution.

With the two-class system, we are no longer in the utopia BhuVai. The two classes of people have two different sets of $\alpha$, $\beta$, and $\gamma$, indicating two different payoff preferences. Hence, they will enjoy different levels of happiness—they may be close, but they will *not* be the same. For want of a better label, let's call this society, as a two-class utopia, a two-class BhuVai. It is an imperfect utopia, if you will. The happiness of all the people in class-1 is the same for everyone in that class—let's call this happiness-1. Similarly, everyone in class-2 enjoys the same level of happiness, happiness-2, which is greater than happiness-1. It is not a perfect society, but pragmatically this may be acceptable if the difference is not too large.

If we cannot have one-class utopia in practice, the next best thing to do is to have such a two-class society where the lower class (i.e., class-1) encompasses an overwhelming majority of the population (say, 99%) with a near-ideal share of the income distribution, and a small minority (say 1%) in the upper class enjoying more happiness, but the economic happiness levels enjoyed by both classes are quite close. We argue that the Scandinavian countries are perhaps close to this two-class utopia.

Similarly, one can create a three-class utopia and so on. In the limit, every agent has a different payoff preference, i.e., different $\alpha$, $\beta$, and $\gamma$ values. This will result in every agent enjoying a different level of happiness, with a large gap between the lowest level of happiness and the highest level. We are no longer in any sort of utopia! It is quite the opposite, in fact! As noted in chapter 4, it doesn't make much sense to consider anything beyond a four-class society as it will be difficult to verify the model's predictions with empirical data.

PAYOFF PREFERENCES AND PERSONAL ATTRIBUTES The parameters $\alpha$, $\beta$, and $\gamma$ model payoff preferences that depend on a number of personal attributes such as motivation, aspiration, talent, qualification, experience, and diligence, which make one successful in one's career. In a highly simplified and aggregated fashion, these parameters model the Maslow-like behavioral and personality needs and traits of the agents. For example, a higher $\alpha$ signals that the agent is highly motivated by salary and that she derives a higher utility from it than an agent with a lower $\alpha$. Likewise, a lower $\beta$ implies one is more diligent and is less bothered by the demands of the job, because the job's disutility is weighted less by this agent. This also implies that this agent is quite likely to be very skilled at her job, and/or derives a lot of pleasure from it, and so the job doesn't seem like a lot of effort, particularly compared with someone with a higher $\beta$ for the same job. Higher $\gamma$ means one is highly motivated by future career prospects.

We assume, in general, that $\alpha$, $\beta$, and $\gamma$ can vary depending on the career or life choice one makes. For someone whose natural inclination is to be an artist, were she to be forced to become a doctor due to, say, parental pressure, her $\beta$ would be high because the effort for becoming and being a physician would seem too great given her tendencies. So her disutility of effort could be so high that her net happiness (i.e., effective utility) is low, despite earning a higher salary as a physician and the rosy future career prospects she has. In other words, she hates the job so much that she is quite unhappy living that life. This will be particularly bad if she doesn't care that much about a high salary or her future prospects, i.e., she has a low $\alpha$ and $\gamma$. Thus, she would have been better off as an artist, from the net happiness perspective, even though she might have earned a lower salary with shaky future prospects. She would have been happier. In the extreme case, when one is simply incapable of doing a particular job due to lack of interest, education, skills, and/or experience (e.g., a high school dropout being considered for the position of a surgeon), the $\beta$ would be infinity for that job, making such a person unqualified for that position.

When people are pursuing what they would like to do and/or are naturally good at, meeting or exceeding expectations, their $\beta$ will be low because the job seems so easy for them. They love doing it. If one is not living one's desired life, then $\gamma$, which weights the future prospects, can be so high that the future prospects look bleak.

When all people are doing the jobs they love, they are at their peak performance. At that time, given their education, experience, effort, enjoyment, and expectations for the future, if you were to ask someone if he would trade his job for another one, he would say no, as he is already at his peak performance and at peak enjoyment derived from work. Therefore, they are all in the same state, and so all their $\alpha$, $\beta$, and $\gamma$ values are the same. We call this property the *principle of equivalence*. In other words, when all people are pursuing their desired lives, then they all weigh the contributions from salary, effort and future prospects the same in BhuVai. Therefore, $\alpha$, $\beta$, and $\gamma$ are the same for everyone in this one-class society.

MOTIVATION FOR CAREER ADVANCEMENT One might ask why anyone would want to be promoted if the effective utility is the same for all agents in the distribution. For instance, if this person is a fresh graduate from college who is hired at an entry-level position in a company in BhuVai, the one-class hybrid utopia, he is just as happy as the CEO of the corporation. So where is the motivation for him to seek career advancement?

When you start off at an entry-level position, you are happy to get a job and a salary, and you are excited about your future career prospects. You are learning a lot and are working at the peak of your abilities at that point, which are, of course, limited as you are new and still learning. You find the job meaningful, challenging, and engaging. And so you are happy.

But after a few years in the same job, you know what to do in the job and how to do it well. You are not learning anything new anymore, and you are operating at *less than peak ability*, not using all your qualifications and abilities. You are bored with the job; it is no longer as challenging and meaningful. As a result, $\beta$ increases along the lines discussed above. You find it harder to get out of bed in the morning to go to work. It seems like so much more effort than it used to be. So the disutility $v$ from the job increases. This also affects the perception of future prospects. You feel you are stuck in this "dead-end" job with no possibility of upward mobility. Hence, $\gamma$ increases, motivating you to look for better future prospects. So the effective utility $h$ drops, and you are no longer as happy as you used to be, which motivates you to search for a new job to increase your utility.

As a result of this search, let us assume you found a new job that you like. You are back to peak performance—back to peak $\alpha$, $\beta$, and $\gamma$—and hence back to peak effective utility—that is, until this cycle is repeated again in a few years, when the time comes for the next promotion. When one gets promoted and vacates a position, another aspiring agent is hired into it. Thus, despite all the stochastic dynamics, the equilibrium distribution is maintained. At equilibrium, every agent is utilizing one's abilities maximally and is at the very peak of performance. Each one is at the peak $\alpha$, $\beta$, and $\gamma$, which is the same for everyone in BhuVai.

A similar behavior occurs when one is treated unfairly at work. Say that you are actually quite happy with your job, until one day when you find out that your colleague earns a greater salary than you do for making essentially the same contribution or maybe even less in your assessment. Or you are passed over for a promotion that you think you deserve more than the person who got promoted. You are naturally upset by this unfair treatment. This negatively affects your perception of your job and of future prospects in your organization. As a result, your preference parameters $\alpha$, $\beta$, and $\gamma$ change along the lines described above, resulting in decreased effective utility. This motivates you to look for a new job where you could be back at your peak parameters, enjoying a higher utility.

It should be the goal of any organized system, whether it is a corporation or society at large, to provide the necessary tools and support so that its members can contribute effectively by performing happily at the peak of their abilities. When this happens, everyone is equally happy, as in BhuVai.

THREE KINDS OF FAIRNESS Furthermore, the effective utility function models three kinds of fairness. First, we have what we call as *intracategory* fairness (Venkatasubramanian 2010). This specifies that all employees in a given category receive the same salary—this enforces the "equal pay for equal work" fairness principle. All employees get the same salary $S_i$ in level $i$. This is the fair recognition of the *present* contributions by an employee.

Second is *intercategory* fairness, where one gets more pay for greater contributions, another founding principle of fairness. This is enforced by having different salaries for different job categories. This is the fairness in compensation for past investments an employee has made in education, training, experience, skills acquired, etc. This is the fair recognition of the *past* investments an employee had made, by employing her at an appropriate job category commensurate with her qualifications.

Third is the utility derived by having a *fair opportunity* at better career prospects in the future. This is the potential for upward mobility. Consider a freshly minted lawyer who has two job offers for an associate position from two different law firms. Both firms are equally prestigious, and both offer the same salary package and workload. But one firm informs our candidate that she will have an opportunity to make partner in eight years if she performs well, while the other doesn't make any such offer. Which offer she is more likely to accept? The one with the partnership prospects, of course. In fact, she might take up that offer even if the salary was somewhat lower and the workload higher because of the future prospect of becoming a partner. Future career (and hence life) prospects are always in our minds when we invest our time, effort, and money in getting an education, acquiring work experience, etc. Thus, the utility of having a fair shot at a better future is an important criterion in our decision-making. This is the utility term $w$. The first two kinds of fairness are represented jointly in the net utility $u - v$. Thus, the utility function models fairness in compensation for the past, present, and future aspects of an employee's professional life.

In summary, our utility function is simply the mathematical representation of the following property of all agents:

All employees expect, at the very least, a fair compensation for their contributions and a fair opportunity for a better future from their jobs that they find to be meaningful, challenging, and gratifying.

## 8.5 Can a Simple Model Handle Free-Market's Complexity?

One might ask[4] how something as complex as a free market—with millions of employees and thousands of companies, buying and selling countless goods and services, worth trillions of dollars, involving myriad job categories that require an enormous variety of talents and skills—could possibly be modeled using a relatively simple model that ignores all these details and complexity.

How could such a complex setup be reduced to essentially five equations concerning (1) the utility function of an agent, (2) the potential function, (3) the maximization of the potential to reach Nash equilibrium, (4) the maximization of total effective utility to get a socially optimal outcome, and (5) Lyapunov stability, with expectations that this simple setup will answer one of the central and vexing questions in economics over the last 200+ years, since the days of Adam Smith and his *Wealth of Nations*: What is the fairest distribution of income in a free-market society? (In fact, to answer this question we need only the first three equations; the last two answer the social optimality and stability questions.)

This is certainly an important question which deserves an answer that hopefully reveals the key insights that explain why our theory makes sense. There are five reasons why we think our modeling is reasonable, as we explain below. Some expert economists might feel the examples below might be a bit of overkill. However, since this question was raised by an expert economist, we feel it is worthwhile to consider such a detailed scenario.

COST-BENEFIT TRADE-OFFS: HOME-BUYING EXAMPLE Consider a typical family of four: Will, Diane, and their two girls, in the market to buy their first home in a midwestern U.S. city. Will just started as an associate at a law firm, and Diane is a dental assistant. The older daughter is in middle school and the younger one in kindergarten. Will makes $100K per year and Diane earns $30K per year.

Will and Diane have their own preferences with respect to the home style (colonial, ranch, split-level, etc.), number of rooms, sizes of rooms, number of bathrooms, kitchen, basement (finished or not), large front and back yards, two-car garage, master suite, neighborhood school district, commute to work and shopping, and so on. They are hoping to spend about $350K, but if pushed, they could go up to $400K, which is a hard constraint. The kids have their own preferences too.

They have seen about twenty homes and are trying to narrow it down to three they'd like to bid on. Now, we could try to write down complicated utility functions that account for the utility derived from each one of these factors of preference for both Will and Diane, compute the utilities, and the marginal utilities of various features, for all the homes they have seen, and try to maximize total utility. However, as we know, this is pointless. First, this is virtually impossible to do as we don't know how to model the functional forms for these factors and/or weight the different factors in the model. Second, no one calculates or evaluates such utilities and marginal utilities in real life as they decide to buy a home.

Instead, in real life, people have a sense of their priorities, some of which act as hard constraints, such as the price of the house, school district, and commute to work, and others act as soft or flexible constraints, such as a finished basement or ranch style. The final decision is arrived at by weighing all these factors intuitively and reaching a compromise allowed by their constraints.

Finally, it came down to selecting from three top choices: (1) a large four-bedroom (both would like a four-bedroom home), three-bath ranch house (Diane prefers ranch), finished basement (Will likes this), with a nice backyard, in a moderately strong school district, with a 45-minute commute each way for both Will and Diane, at $350K; (2) a smaller three-bedroom, two-bath colonial, in a very strong school district, with a 20-minute commute for both, at $425K; and (3) a large four-bedroom, 3.5-bath ranch house, finished basement, large front and back yards, in a very strong school district, with a 15-minute commute each way for both, at $550K.

Will and Diane really loved the third home, as it pretty much had all the features they were looking for, but it was pricey. They tried to negotiate the price down, but they could only get it down to $500K and were priced out. The first home also satisfied a number of their criteria, but they were not happy about the long commute or the fact that it was not in a very strong school district. However, they were willing to buy it, but not at $350K, even though they could afford to. They offered $300K, but the seller refused to come down below $340K.

So they ended up compromising and buying the second home, even though it did not have a number of the features they wanted. It was a smaller home, not a ranch, no finished basement, etc., but with a short commute and a strong school district. For these features, they were willing to sacrifice some other things they wanted and to pay a higher price than they were planning to spend. They negotiated and brought the price down from $425K to $400K.

A detailed model of the overall utility function (and the associated marginal utilities) for all three houses, were it possible to develop, would account for the complexities of the situation, by considering the utility Will would enjoy from a four-bedroom, three-bath, home with a nice front and back yard and finished basement (by appropriately weighting all these factors), and subtracting the disutility caused by a weak school district and longer commute (again, suitably weighted). Likewise, this is done for Diane, who has her own utility function that is different from Will's. This can be done even for the kids, if their preferences are taken into consideration. However, it is practically impossible to perform all this.

In reality, however, millions of people make these assessments and judgments all the time, when they are buying (or renting) a house. They do this by intuitively weighing all these pros and cons, utilities and disutilities, and essentially converting all this into a *single quantity* that captures the overall cost-benefit trade-off they are willing to make, namely, the *price* they are willing to pay.

Likewise, in determining the price of the house, one could develop a detailed supply-demand model of the houses in the neighborhoods Diane and Will are interested in and then calculate the appropriate price. Using the recent sales data, one could presumably develop a model that accounts for a wide variety of variables and parameters such as the area of the land/house, number of bedrooms, and bathrooms, condition of the property, etc., and then determine the price of the properties they are interested in. While neither the seller nor the buyer has perfect information on the current supply and demand forces, the recent sales data often provide a reasonable picture of the market.

Typically, people don't use these data to develop detailed supply-demand models. Even if one could develop such models, it is not even clear how reliable and useful they would be, given the errors and uncertainties involved in practice. People instead form *semiquantitative* impressions of the market, guided by real estate agents, and make their own assessments about the bid-offer prices of various properties. Despite this guidance, the buyer typically does not know precisely how low she can bid. Likewise, the seller does not know exactly how high he can list the price. Nevertheless, buyer and seller negotiate to find a compromise. If the deal goes through, then it implies that the supply and demand forces have balanced each other, at least locally. Thus, if the market clears, then the tension between supply and demand has been *reflected* and *resolved* in the sale price. If the offer price is too high for the current demand (or supply), then the price has to come down for the market to clear. If the bid is too low for the current demand (or supply), it has to increase to close the deal. Thus, even without the help of detailed supply-demand models, millions of people buy and sell homes every year, guided by recent sales data, because everything one wants to know about the local supply and demand forces is already reflected in the *sale price*.

The point of this analysis is that despite all the complexities in computing the overall utility and marginal utilities of houses, and the supply-demand characteristics of the local real estate market, everything is finally reduced to one crucial quantity, the price, which accounts for all this. And at the right price, the *sale price*, the market clears, and the transaction takes place. So while there are so many factors, in principle, that affect the outcome, in the end, the only factor that matters really is the price. Everything has been accounted for in the sale price, for both the buyer and the seller. For Diane and Will, the first house made sense at $300K, but not at $340K. Likewise, they were willing to spend more than their original amount, at their upper bound at $400K, for the second house. Thus, the prices in both cases had captured the overall utility-disutility, benefit-cost trade-offs for them as well as the overall supply-demand profile of the market.

COST-BENEFIT TRADE-OFFS: JOB-OFFERS EXAMPLE Now consider a new scenario, the next development in their lives. After living two years in their first home, Will gets two job offers as senior associate, one in the same law firm where he currently works and another at a prestigious law firm in a big city in the southeastern United States. The total compensation for the first offer is $130K per year, and the second is $160K per year. The southeast law firm has also agreed to help find a suitable job for Diane and pay for the relocation expenses.

Once again, as Will and Diane evaluate the pros and cons of both offers, they realize that the overall utility is a complicated aggregate that depends on a host of factors, some measurable, others not. Obviously, the overall utility is dominated by pay (i.e., total compensation including base salary, bonus, options, etc.), but it is also dependent on other factors such as quantity (e.g., hours of work, work-related travel away from home) and quality (e.g., interesting or boring) of the work in the new job, title and peer recognition, competition and job security, career and personal growth opportunities, retirement and health benefits, company culture and work environment, disruption due to relocation, and so on, not necessarily in that order.

In particular, for the second offer, there are disutilities associated with the potential disruption to their lives due to relocation to a new and unfamiliar city. They will have to buy a new home (and go through, once again, the agony they went through a few years back), find new schools for their kids, try to form a new social network of friends, etc. While new and exciting benefits might arise in the new environment, they are not as certain as the loss they will suffer from leaving a familiar and comfortable life and environment. The disutilities of such a disruptive move are quite clear and certain, while the potential benefits are unclear and less certain. However, the offer is at a prestigious law firm, and they have offered to preselect Will for a fast track to partnership in four years. While the partnership is not a given, Will certainly has less competition as he is already one of the select few on a fast track. This partnership is also potentially worth more monetarily over his career life than the one from his first offer.

The first offer, however, will be less disruptive as they can continue to live in the same home, send the kids to the same schools, and enjoy their current social life. Diane can continue to work at her dental office, which she enjoys. While there are some uncertainties associated with the new position, Will already knows a lot about his current law firm, minimizing the potential for nasty surprises. In his new position, he will be considered for partnership in six years.

As with the house purchase example considered above, it is impossible to write down the model for the overall utility (and marginal utilities) that accounts for all the utility or disutility of various factors, such as quantity and quality of work, peer recognition, retirement and health benefits, company culture, disruption from relocation, and so on. Likewise, a detailed supply-demand model of lawyers is fraught with its own difficulties.

Again, Will and Diane weigh intuitively the pros and cons of both offers and conclude that the second offer is worth pursuing at $200K, but not at $160K. They figured that extra $40K per year compensates for the disutility due to relocation. Thus, salary acts as a proxy for the disutility associated with the job. They also figured the prospect of partnership at a prestigious law firm also compensates for the disruption and uncertainty. Thus, they reasoned to compensate for the disutility by taking part of the compensation as cash now (as in the extra $40K per year) and the rest as better future prospects. So Will negotiates for a higher salary, and happily the southeast firm agrees. Diane finds a new job at $35K, and they relocate to their new jobs.

Once again, as in the case of the house purchase, all the complexities of the competing utility and disutility factors, and the supply-demand tension, are finally reduced to a single quantity, namely, the *salary*. At $160K, there was no deal, but at $200K the deal went through and the market cleared.

Again, the point of this analysis is to show that millions of people make these kinds of decisions all the time, by reducing and expressing the complex trade-offs to a single dominant factor—the *price* of the object of interest. Millions of people routinely make such comparisons of disparate entities or features, going way beyond apples-to-oranges comparisons, and somehow condense all the complexities to a single quantity of interest. If the object is a house (or some such thing), it is the sale price of the house. If it is a job, it is the price of the job, namely, the salary at which the deal is executed. This is why, we think, it is reasonable to express the overall utility function in terms of the single dominant quantity, the salary. This is the first insight, the first reason.

We also note that the price of a house *directly* specifies the *cost* of acquisition (i.e., its disutility), while it *indirectly* signals the *benefits* (i.e., its utility) one might enjoy living in it. One can be reasonably certain that a million-dollar house would provide a lot more benefits than a $300K house in the same neighborhood. The cost (i.e., disutility) is explicitly displayed, but the benefits (i.e., utility) are implicit. In contrast, the salary of a job directly specifies the benefits, while it indirectly signals the cost of getting and keeping the job. A higher-paying position typically will demand a lot more from an employee than a lower-paying one. Thus, both price and salary convey information about the cost and benefit of the entity of interest.

COMMON FEATURES OF THE UTILITY FUNCTION The second reason for our modeling framework is the insight that most cost-benefit, disutility-utility, trade-off functions in real-life have the inverted-U profile, as discussed in section 3.2. This feature covers a lot of decision-making scenarios, thereby justifying the functional form we have in equation (3.6).

The third reason is the insight that for most people the overall utility of a job is derived from two main components: (1) the immediate net utility of making a living from the salary earned and (2) the utility of a better future life that the current job could lead to, as modeled in equation (3.6).

The fourth reason regards the preference parameters $\alpha$, $\beta$ and $\gamma$. Based on one's talents, skills, education, experience, effort, accomplishments, and ambitions, these parameters can vary from person to person. However, as discussed in section 8.4, in our utopia, when people are pursuing what they would like to, working in a job they love doing, meeting or exceeding expectations, they are as happy as they can be from work. They cannot think of themselves doing any other job. When everyone is in this state, by symmetry, everyone has the same preference for the three utility components $u$, $v$, and $w$. The utility they derive from salary, discounted by the effort they have put in, with the prospects of a bright future are all weighted the same way for everyone in BhuVai. Everyone is in her ideal state. When all employees are at their ideal jobs, then they all have the same parameter values. Therefore, $\alpha$, $\beta$, and $\gamma$ are the same for everyone. This is the fourth insight.

Finally, the millions of employees and thousands of companies and job categories actually help rather than hurt the modeling, because of the benefits of statistical averaging, for example, as occurs in the insurance business.

We believe that while our model is simple, it is not simplistic. While it ignores many details, it captures the essential features, as discussed above. Thus, these four key insights, which are generic and hence widely applicable, coupled with the statistical benefits of a large population of agents, justify why our modeling framework could be expected to work.

PRINCIPLE OF MAXIMUM ENTROPY: EARLY OBJECTIONS It is both interesting and important to note that similar objections were raised regarding the simplicity of the approach, given the complexity of the problem, in the context of statistical thermodynamics as well. As Jaynes writes,

> From Boltzmann's reasoning, then, we get a very unexpected and nontrivial dynamical prediction by an analysis that, seemingly, ignores the dynamics altogether! This is only the first of many such examples where it appears that we are "getting something for nothing," the answer coming too easily to believe. Poincaré, in his essays on "Science and Method" felt this paradox very keenly, and wondered how by exploiting our ignorance we can make correct predictions in a few lines of calculation, that would be quite impossible to obtain if we attempted a detailed calculation of the $10^{23}$ individual trajectories.
>
> It requires very deep thought to understand why we are not, in this argument and others to come, getting something for nothing. In fact, Boltzmann's argument does take the dynamics into account, but in a very efficient manner. Information about the dynamics entered his equations at two places: (1) the conservation of total energy; and (2) the fact that he defined his cells in terms of phase volume, which is conserved in the dynamical motion (Liouville's theorem). The fact that this was enough to predict the correct spatial and velocity distribution of the molecules shows that the millions of intricate dynamical details that were not taken into account, were actually irrelevant to the predictions, and would have canceled out anyway, if he had taken the trouble to calculate them (Jaynes 1979, 16–17).

In a similar manner, the details we left out in our model are irrelevant to the questions we are asking, as we argued above.

Furthermore, as we discussed in sections 6.2, 8.1, and 8.4, the equality of the preference parameters corresponds to the one-class utopian society. We *do not need* to have this realized in a real-world capitalist society for this result to be useful. This ideal state, which we *do not expect* to see in real life, is still relevant and useful, because it defines the best possible outcome of a capitalist society, the fairest distribution of income, and the associated social optimality and moral justification. As noted, knowing this reference state, like the bull's-eye in a dartboard, helps us measure how far the real-life societies have deviated from this ideal world, and helps us formulate targeted policies to correct for unfair inequalities.

## 8.6 What About Economic Growth?

What does our theory have to say about economic growth? The theory offers guidance about how the income pie ought be shared among the population. But what about growing the pie? Since our theory is about fairness and not growth, it does not have much to say about growth directly, but it does have growth-related implications, as we discussed in section 6.4.

The general sentiment in mainstream economics, as reflected by Robert Lucas's statement (see chapter 1), is that growth is much more important than fairness. The sense is that it is better to have a larger pie inequitably shared than to have a smaller pie that is shared more equitably, as long as people at the bottom are getting more in absolute terms, even though they may be receiving less in relative terms. This is certainly true in extreme cases, but it is not as straightforward as one might imagine when one considers other scenarios.

Consider, e.g., the per capita GDP of India and of the United States in 2014 (in 2014 dollars): $\sim$$1,500 (India) and $\sim$$54,000 (United States). It is, of course, true that an average citizen is better off in the United States than in India due to this extreme disparity in per capita GDP (India also has a high income inequality, which makes this even worse). Yes, the absolute level of income (as signaled by the per capita GDP) matters a lot.

Let us now consider another situation. Consider the following two scenarios: (1) per capita GDP is $\sim$$20,000 and $\psi_{90} = -12\%$ (i.e., the bottom 90% is receiving 12% less than its fair share of income) and (2) per capita GDP is ~$50,000 and $\psi_{90} = -24\%$. Which scenario would be better for an average citizen?

The first scenario is less unfair, but the per capita GDP is low, whereas the second scenario is certainly more unfair, but the per capita GDP is much higher. This might suggest that the latter is perhaps better for an average citizen. The overall economy is doing so much better—in a sense, the income pie is much larger. Obviously, the bottom 90% is much better off in the second rather than in the first scenario, right?

Well, not exactly. These are not fictitious numbers, but actual data. The first scenario corresponds to the United States in 1965, and the second is the United States in 2014 (per capita GDP is in constant 2009 dollars). We all know that the bottom 90%, the average citizen, and the middle class were all much better off in 1965 than they were in 2014 (or now for that matter). And this was not just an isolated result, true for only 1965. It was true for roughly a 30-year stretch, for almost two generations, from 1945 to 1975. On the other hand, the middle class and the bottom 90% have largely stagnated for the last 35 years, from 1979 to 2014.

Thus, the argument that everything will be fine if we grow the pie without worrying about equity is not quite correct. This reasoning is perhaps valid for extreme cases, such as India versus the United States, where the per capita income disparity is dramatically high (U.S. per capita GDP is about 35 times that of the Indian per capita), but not for more moderate cases such as the one we noted above. In extreme cases, the dramatic increase in the absolute income level more than compensates for the negative effects of income inequality. But when the increase is more modest (it is about 2.5 times between 1965 and 2014 in the example above), the damage done by the sharp increase in income inequality (the bottom 90% went from losing ~12% of its fair share in 1965 to losing ~24% of its fair share in 2014, in the U.S.) is not so easily compensated for.

There are several reasons for this, which we already discussed in chapter 1. They all result from the widening gulf of affordability between the haves and have-nots. The top 10% benefit so much more from the growth of the pie that their affordability and affluence drive up the price of access to better quality of services such as health care, housing, education, and justice. Furthermore, their economic power translates into political power, which they exercise to advance their parochial interests, such as cutting taxes for the wealthy, curtailing services for the bottom 90%, and so on. As a result of all this, the bottom 90% gets basically priced out of a decent quality of life.

There is another vital factor to consider from the growth perspective. If one analyzed the last 100+ year history of innovations in the United States, one would find that an overwhelming majority of innovators came from the middle to lower economic strata of the society, i.e., from the bottom 90%, and not that many from the top 10%. Certainly, negligible numbers came from the top 1%. The stratum that has contributed the most is the middle 60%, in the typical range of about $45K to $125K (2015 dollars) in annual household income. It is very troubling that this stratum has been doing poorly with respect to its economic health in recent years. This, as we argued, negatively impacts other important metrics such as its access to education, health care, etc., which doesn't bode well for our continued dominance in technology.

To ensure U.S. leadership in technology and innovation, which is the bedrock of our economic growth and prowess, we need to take care of this economic stratum particularly well. Putting moral arguments aside, at least on the basis of a purely utilitarian, self-interest, perspective, we need to support this segment particularly well. They are the engine of our future growth, requiring our great care and attention.

## 8.7 Comparison with Econophysics

As noted by Kirman (2011) and Philip Mirowski (1989), historically concepts from physics and researchers trained in physics have played an important role in shaping our thinking in economics. Concepts such as equilibrium, forces of supply and demand, and elasticity reveal the influence of classical mechanics on economics. The analytical model of utility-based preferences can be traced back to Daniel Bernoulli, the great Swiss mathematical physicist from the eighteenth century. One of the founders of neoclassical economics, Irving Fisher, was trained under the legendary Yale physicist, Josiah Willard Gibbs, a cofounder of the discipline of statistical mechanics. Similarly, Jan Tinbergen, who shared the first Nobel Prize in Economics in 1969, was the doctoral student of the great physicist Paul Ehrenfest at Leiden University. Thus, looking to physics and physicists is nothing new or weird in the development of economic ideas, even though, as we discussed in section 5.10, this has not been the case in recent decades.

In the 1990s, physicists trained in statistical mechanics started applying their techniques to problems in economics and finance. The term *econophysics* was coined by the Boston University physicist H. Eugene Stanley to describe this emerging discipline. One of their main contributions has been on the topic of income and wealth distributions using thermodynamic models. However, as noted in section 1.4, they have not adequately addressed the conceptual gulf that exists between econophysics and economics in two critical areas: their "zero intelligence" particle models of agents and the interpretation of entropy as a measure of disorder and uncertainty.

Furthermore, most importantly, they have not addressed the issue of fairness in their theories and models, or the related issues of moral justification and stability of a free-market capitalist society.

Even though our contribution also utilizes concepts from statistical mechanics, it takes an entirely different perspective by addressing these conceptual challenges directly—particularly by tackling the *vital issue* here, namely, fairness. There has been considerable work done by political philosophers, and to some extent by economists, on fairness (as discussed in chapter 2). But these approaches have not addressed whether the free-market dynamics will lead to a fair distribution (beyond what Nozick asserted; but he did not forecast what kind of distribution will emerge). The question of which of various measures of fairness is most appropriate in a free-market environment has been neither raised nor answered before (Venkatasubramanian 2010). Indeed, conventional wisdom in economics is that the free market cares only about efficiency and not about fairness. Thus, there is a disconnect between the econophysics and mainstream economics–political philosophy communities in this context. The former has proposed models inspired by statistical mechanical analog but has not interpreted entropy in economically relevant terms. In particular, it has not addressed the issue of fairness in its theories. In contrast, the latter, which has proposed many theories of fairness, has not recognized the relevance of, and connected with, the statistical mechanical theories. Our theory bridges this gulf, by identifying the deep connections between these two approaches via the concept of entropy as fairness.

Furthermore, econophysicists typically (Chatterjee et al. 2005; Yakovenko and Rosser Jr. 2009; Yakovenko 2012; Richmond et al. 2013; Chakrabarti et al. 2013) like to claim that the bottom ~95% follows the Boltzmann-Gibbs (BG) exponential distribution or a gamma distribution, not lognormal, and that the top ~3%–5% follows a Pareto distribution. We beg to differ on both counts as we have shown in our work. One main difficulty with the BG exponential or the gamma distribution claim is the interpretation of the underlying economic notions. For example, from the maximum entropy procedure that underlies these claims, we can show that the BG exponential distribution implies a utility function that is linear in salary (Kapur 1989) which conflicts with the principle of diminishing marginal utility, one of the fundamental principles in economic theory. At the risk of repeating ourselves, we emphasize that in our framework we have formulated an approach that is sensible from a microeconomic perspective, e.g., modeling agents with reasonable utility preferences, rational agents making decisions motivated by utility maximization and not due to random events, recognizing entropy as fairness, etc.

Similarly, there has been some work in the past that explored the connection between game theory and statistical mechanics (Blume 1993; Wolpert 2006; Kikkawa 2009). What is new about our theory is that it shows a direct and deep connection between the dynamics of animate, fairness-driven, utility-maximizing, rational *teleological* agents and inanimate, purpose-free, thermally driven molecular entities. Our result reveals the surprising and important connection between entropy and game-theoretic potential, demonstrating that the statistical thermodynamic equilibrium reached by molecules is really a Nash equilibrium. We believe that this is a significant insight, for it suggests that statistical thermodynamics can be seen as a *special case* of potential game theory. Alternatively, one may view this insight as the generalization of the laws of statistical thermodynamics to teleological systems, such as economic systems, yielding a new conceptual framework, which we call *statistical teleodynamics*, that unifies statistical thermodynamics and population game theory. This framework bridges the conceptual gulf mentioned above, as our ideal teleological agents are rational, fairness-seeking, utility-maximizing strategists, but with a natural connection to statistical thermodynamics.

Another important related observation is that, in statistical thermodynamics, the claim about the equilibrium state is a *probabilistic* one—it is the most probable outcome, one where entropy is maximized. However, our game-theoretic result shows that the Nash equilibrium state reached by the molecules, the one that maximizes the potential $\phi(\mathbf{x})$ is a *deterministic* outcome, not a probabilistic one. This observation has potentially important implications concerning the philosophical foundations of statistical thermodynamics, and those of information theory, such as ergodicity and metric transitivity (Tolman 1938; Khinchin 1957; Theil 1967; Jaynes 1979; Sklar 1995; Reif 1965; Nash 2012), but we do not address them here.

Besides the conceptual challenges, there is a technical one due to the nature of the data sets in economics. As Ormerod (2010) and Perline (2005) discuss, one can easily misinterpret data from lognormal distributions, particularly from truncated data sets, as inverse power law or other distributions. Therefore, empirical verification of econophysics models is still in the early stages.

## 8.8 Future Directions: Theoretical and Empirical Studies

There are obvious limitations to our model—we have assumed perfectly rational agents, no externalities, ideal free-market conditions, and so on, which are clearly not valid in real life. However, our objective was to develop a general microeconomic framework, identify key principles, and make predictions that are not restricted by market-specific details and nuances. Despite such simplifying assumptions, it is encouraging that our predictions are supported by empirical data. Additional work is needed to develop and test our theory more rigorously and extend it for more realistic market conditions.

THEORETICAL AND AGENT-BASED SIMULATION STUDIES The following is a partial list of research topics one should investigate by developing the theoretical framework further. These theoretical extensions should be carried out simultaneously with agent-based simulation studies as well as with empirical testing and validation.

- What is the equilibrium *wealth* distribution in BhuVai?
- Study the effect of including savings and unearned income such as dividends in the income distribution.
- Examine two-class, three-class, and four-class societies in detail. In particular, examine the case where $\alpha$, $\beta$, and $\gamma$ are drawn from a probability distribution for each class. That is, for the two-class system, for instance, instead of assuming all class-1 agents have the same $\alpha$, $\beta$, and $\gamma$, assign them values drawn from a probability

distribution with a given mean and variance. Do similarly for class-2, and then study the equilibrium properties.
- Examine our assumption that approximates an entire country as a large corporation functioning in a free-market environment for the country-wide comparisons in chapter 6. Can we model this more rigorously?
- How do we model market power and rent seeking by the executive class? On the flip side, how do we model the loss of bargaining power by the working class?
- How do we generalize our uniqueness, stability, and optimality results, obtained for labor markets, to the economy as a whole and contrast them with the Walrasian general equilibrium model. What are the lessons to be learned here?
- Extensive theoretical, simulation, and empirical studies are needed to study the stability and resilience of real-life free markets. How much unfairness can be tolerated before we lose the stability properties of free-market dynamics?
- Examine how concepts, techniques, and lessons from process systems engineering could be adapted and applied to analyze free-market societies, particularly concerning the modeling, control, optimization, and risk management of nonlinear dynamical systems (see, e.g., Bookstaber et al. 2015).
- Our model assumes that a continuum of skills, from blue-collar skills to executive-class skills, are in demand in the market. What if the demand for some skills reduces and that for others increases unevenly? To some extent, this is what has happened in recent years with the demand for the blue-collar skills decreasing as a result of globalization, while the demand for the high-tech worker skills rises with the emergence of the digital economy. How do we model this? How will this change the equilibrium income distribution?
- With the advent of applied artificial intelligence, it is quite likely that what happened to blue-collar skills in $\sim$1980 to 2010 due to globalization is likely to happen to white-collar skills in $\sim$2015 to 2040. There would be a hollowing out of the demand in the human skills spectrum, from the low end through the middle up to the near upper end. This could result in a winner-take-all economy, much more than what is seen now. What would be the effect of this on income and wealth inequality and societal stability?

- One important trend we are beginning to see is the emergence of the "gig" economy. A gig economy is an economic environment in which temporary positions are more prevalent and organizations (or even people) contract with independent workers for short-term engagements or gigs. In such an environment, more people choose to make their living by working gigs than by working full time in regular jobs that pay a monthly salary. If the gig economy became the norm in the coming decades as a result of technological progress, what would the income and wealth distributions look like? Would it increase or decrease income and wealth inequality?
- We showed in chapter 5 how a lognormal distribution emerges when we maximize fairness under the constraints on mean and variance. What if, in addition, we have other information, such as reliable metrics on the performance of employees? How do we formally incorporate this information in our theory to determine the fairest salary distribution?
- As we noted in chapter 7, a new analytic theory of value creation in a large corporation is needed. As a product emerges from an idea to a marketed product, as it goes through various stages of its creation, as the employees add more and more value in these stages, what does the value trajectory look like for various classes of products such as commodities, specialty items, software, hardware, intellectual products, pharmaceuticals, and so on? How do we develop quantitative metrics to track this value trajectory and tie it to the pay packages of all the employees who created it? If we had such a value model, how would we use it in our framework?
- Study the effects of peer network, social capital, and other externalities.
- One important topic we don't address is questions that arise in a society in which some citizens are disabled, one way or the other, requiring special treatment. These questions have been addressed by Amartya Sen, Martha Nussbaum, and others in their frameworks. Our theory, in its present form, is focused on only able individuals who generally constitute the bulk of a population. Again, this is another natural extension to investigate.
- One challenging theoretical issue is whether the ideal free market needs to be ergodic for our theory to be applicable. Jaynes (1979) asserts that his formulation of the principle of maximum

entropy dispenses with the requirements of ergodicity, metric transitivity, etc. However, this is not a settled issue, as observed by Sklar (1995). This question needs to be explored further for free-market dynamics.

EMPIRICAL STUDIES

- While the empirical agreements with the aggregate model predictions are encouraging, more thorough studies are needed using detailed distribution data to validate these initial impressions and understand the comparisons more rigorously. We need to study, for example, *percentile* and *decile* breakdown of income data and compare them with the theoretical predictions for various countries.
- We need to conduct more comprehensive studies of pay distributions in various organizations in different countries, in order to understand in greater detail the deviations from ideality in the marketplace. We should collect and organize empirical income distribution data to analyze global patterns across several dimensions:

    1. Organization size—small, medium, large, and very large number of employees.
    2. Different industrial sectors, e.g., chemicals, oil and gas, financial, software, hardware, communications, construction, and services.
    3. Different types such as private corporations, governments (state and federal), and nonprofit organizations.

- Further studies are needed to compare the model predictions with post-tax and transfer income data from different countries. In addition to gathering empirical data, it would be good to build a large-scale agent-based simulation program that also accounts for taxes, savings rates, returns on assets, etc. Such a program can help us carry out various "what if" scenarios to test the effects of different tax and transfer policies, in order to arrive at optimal policies that balance growth and inequality.
- The effect of including health care benefits provided by the employer as part of the compensation, particularly for the bottom 90%, needs to studied carefully as this may reduce the perceived income inequality. This is particularly important for U.S. income data.

- As noted in chapter 6, careful studies are needed to identify the most useful nonideal inequality coefficient $\psi$, and compare its performance with that of the Gini coefficient.

In summary, we view our theory only as a start of a new direction of inquiry in distributive justice. Much more remains to be done.

## 8.9 Summary and Final Thoughts

Inequality per se is not bad. As different people make different contributions, some more others less, in a corporation or in a society, it is only fair that those who contribute more are rewarded more. The question is how much more? That is, how much income inequality is fair? Addressing this important question requires a quantitative, analytical, theory of fairness whose predictions can be verified with real world data. This is the challenge we have dealt with in this book. Our unorthodox, transdisciplinary theory integrates foundational principles from disparate disciplines into a unified conceptual framework that includes the key perspectives on this problem—the perspectives of political philosophy, economics, statistical mechanics, information theory, game theory, and systems engineering. This new paradigm is a synthesis of liberty, equality, and fairness principles from political philosophy with utility maximization by rational self-interested agents in economics, which dynamically interact with one another and the environment modeled by entropy and the potential from statistical mechanics, information theory and game theory, subject to the efficiency and robustness criteria of systems engineering.

Our new theory, which we call *statistical teleodynamics*, rests on two key conceptual insights. One is that the concept of entropy from statistical mechanics and information theory is the same as the potential from game theory, and that these mathematically model the concept of fairness in economics and philosophy. The other key insight is that when one maximizes fairness, all agents enjoy the same level of effective utility at equilibrium in an ideal free-market society. These insights help us prove that the fairest inequality of income is a lognormal distribution, at equilibrium, under ideal conditions. One may view our result as an "economic law" in the statistical thermodynamics sense. The ideal free market, guided by Adam Smith's invisible hand, can self-organize to discover and obey this economic law if allowed to function freely without rent seeking, market power exploitation, or other such unfair interferences subverting the free-market mechanism. This result is the *economic equivalent* of the Gibbs-Boltzmann exponential distribution in statistical thermodynamics.

We prove that even though an individual employee cares only about her utility and no one else's, the collective actions of all the employees, combined with the profit-maximizing survival actions of all the companies, in an ideal competitive free-market environment of supply and demand for talent, under resource constraints, lead toward a fairer allocation of income, thereby maximizing social utility, guided by Adam Smith's invisible hand of self-organization. Thus, the ideal free market promotes fairness and social utility, and not just efficiency, as an emergent property. We call this the *fair-market hypothesis*. Since maximizing fairness leads to the equality of economic welfare for all citizens, and maximization of total effective utility, at equilibrium under ideal conditions, our theory provides the moral justification for the free-market economy in analytical, quantitative, terms.

The central principle of maximum fairness, represented by the principle of maximum entropy, helps us design the foundations of a maximally fair capitalist society. It guides us naturally toward the two fundamental principles of Rawls and then consistently combines these with the Nozickian free-market doctrine. It bridges the macro and micro perspectives in an internally consistent conceptual framework. It accounts for macro concepts such as equilibrium and entropy from statistical mechanics, potential from game theory, Lyapunov stability from dynamical systems, and efficiency and robustness from systems engineering.

This is all well and good, you might say, but what can be done about extreme income inequality in real life?

We do not address this important question in terms of economic prescriptions, for it is best left to experts such as Atkinson, Krugman, Piketty, Reich, Stiglitz, and others (Atkinson 2015; Krugman 2014; Piketty and Goldhammer 2014; Reich 2015; Stiglitz 2015), who have written extensively about the cures in their op-eds, papers, and books. For example, Anthony Atkinson, widely regarded as the father of the modern approach to economic inequality, discusses his fifteen-point policy proposal to address the problem in his recent book (Atkinson 2015).

While these experts may differ on the details, they all generally agree that the cure involves an appropriate mix of progressive taxation on the wealthy and corporations, a living minimum wage, universal health care, inheritance tax, improved corporate governance, assistance for education and skills development, and so on. Put simply, we need to restore fairness and upward social mobility by adopting policies that empower all individuals to continue to innovate and grow a free-market economy. In some sense, the cures are all well known. In fact, economists have known them for quite a while. What is lacking is the social conscience and political will to get it done in most countries.

Even though we do not provide policy prescriptions, we believe our theory can still help in an important way. By defining the *best, theoretically possible* hybrid free-market society in quantitative terms, our theory has identified the target to shoot for regarding distributive justice. The target society where everyone is equally happy at equilibrium is the *best* that a society can deliver to its citizens. This is the utopia everyone would love to live in. We can't do any better than this.

We now know that the income distribution in this utopia is the lognormal distribution. Therefore, this is the *ideal benchmark* to shoot for in the real world, which we quantify by using a new measure of unfairness in the income distribution, the nonideal inequality coefficient $\psi$. So if a real-world society comes close to this utopia (i.e., low $\psi$ values), it is doing really well for all citizens, not just the top 10%. Use of $\psi$ could help us better design our tax and transfer policies, executive compensation packages, etc., to improve income inequality by getting as close to ideality as desired.

Our results reveal a surprising outcome. It is the finding that for an overwhelming majority of the population (bottom ~99%), Norway, Sweden, Denmark, and Switzerland (to a lesser extent Netherlands and Australia) have achieved income shares that are close to the values in the utopia. What is even more surprising is that these societies did not know, *a priori*, what the ideal, theoretically fairest, distribution was, and yet they seem to have "discovered" a near-ideal outcome empirically on their own. Based on this result, one could argue that these societies are actually functioning (almost) like the free-market societies envisioned by Adam Smith.

These countries, particularly Scandinavia, are generally derided by free-market enthusiasts as socialist welfare states, far from being free-market capitalist societies, while the United States is generally considered to be the shining example of free-market capitalism. Our theory suggests it is quite the opposite, in the context of distributive justice. The United States is where the ideal free-market mechanism has broken down the worst.[5] While the American free-market democracy has many great things going for it, as evidenced by its ability to attract copious global talent, and the constant stream of innovations produced year after year that lead to economic growth and dominance, we have dropped the ball on distributive justice in the last three decades.

By the way, all six countries mentioned above made it to the top ten happiest countries to live in, out of 150 surveyed, in the 2015 U.N. World Happiness Report.

We don't generally expect to find theoretical predictions to be verified so closely in social sciences. After all, humans are not as predictable as molecules. So naturally, we wouldn't expect the ideal free-market society to exist in real life, given all its simplifying assumptions that are hard to satisfy in practice. That is why it is so surprising to find the Scandinavian societies performing so closely to the ideal in real life for an overwhelming majority (bottom ~99%) over the past two decades. One should take this as encouraging news for the other societies that may strive to accomplish this goal.

Our theory shows how the free market is supposed to work, but in practice, the market doesn't hum along so merrily when left to itself. Every now and then, capitalism seems to run into trouble because of another inalienable part of human nature—naked greed, as Adam Smith warned us in his "vile maxim" remark. Periodically, free-market capitalism seems to get gamed, hijacked by crony capitalists, subverted by rent seekers, aided and abetted by self-serving politicians, to end up as *ersatz capitalism*, as Stiglitz calls it. The concentration of money has led to concentration of economic power and political power, which have distorted and subverted the free-market mechanism. This is the real danger facing us—not socialism or communism.

We consider extreme inequality as a case of *systemic failure* in capitalism in practice. In a number of ways, this failure is similar to the systemic events in other domains involving complex sociotechnical systems such as large chemical plants (e.g., BP oil spill, Bhopal, and Piper Alpha), power grids (e.g., Northeast Power blackout and India Power blackout), and financial institutions (e.g., Global Financial Crisis and Savings and Loan Crisis). As this author has argued (Venkatasubramanian 2011; Venkatasubramanian and Zhang 2016), such systemic failures do not happen just because one or two things went wrong. They occur only when several things fail, and that too over a long period—systemic failures don't happen overnight. The extreme inequality in the United States, and in other countries, as we discussed in chapter 6, is due to several causes, but mainly: (1) weakening of unions, (2) rent seeking by executives, (3) failures in government policies and deregulation, (4) globalization—uneven demand/supply balances in the skills spectrum, (5) failures in corporate governance, (6) a winner-take-all corporate culture enabled by the hero worship of corporate chieftains and hustlers of finance, and their mantra "Greed is Great," and (7) the myopic focus on shareholder value as opposed to stakeholder value. Systemic failures cannot be fixed by changing one or two things. There is no magic bullet. That is the nature of systemic failures. They happen when several things go wrong over a long period. The cure, then, involves doing several things right over time. This means we address the failures listed above and adopt the policy recommendations we discussed a couple pages back. In short, properly addressing this critical challenge requires economic, political, and social reforms.

As before, we need to rescue capitalism from capitalists, or at least from crony capitalists. So what a sensible society needs is to implement actions to counter these innate destabilizing tendencies, to protect the free market. Adam Smith's vision of capitalism had moralistic undercurrents, as we have noted. He believed that capitalism was for the many, not the few—for the advancement of the public good. Capitalism's true power as an incredible engine of innovation and wealth creation is realized only when it is in the hands of many, not just a few. Its genius lies in its ability to tap into the creative potential, industry, and entrepreneurial spirit of all people, not just the top 1%. This means the bottom 90% need to get their fair share of the prosperity they helped to create in the first place, so that they are empowered to innovate and power the future growth of a capitalist society.

Let us pause and take a quick look back at the almost magical progress the Western capitalist societies have accomplished in a mere two hundred years, which is just a blink of time in human history. The technological advances in a number of areas, such as food, clothing, shelter, health, education, communication, and transportation, are simply breathtaking. These societies have empowered hundreds of millions to get out of poverty and enjoy lifestyles that even the kings and queens of Europe did not enjoy two hundred years back. This happened because the vast oppressed majority was finally able to break free from serfdom and channel their talent, innovation, and hard work to become extraordinarily productive.

As history amply demonstrates, the moneyed interests bitterly fought this empowerment all along to protect the status quo which favored them. The battles won to free slaves, to free children, to improve working conditions and hours, to recognize women as equals, and so on, did not come easy. The wealthy, except for a few noble souls here and there, typically fought against all this progress, driven by their "vile maxim," issuing dire forecasts of gloom and doom that the society was going to implode if one didn't listen to them. Instead, as more and more people participated in, and benefited from, free-market capitalism, society only exploded with growth and prosperity. But some still refuse to learn from history, as we continue to hear the same old refrain of how the "takers" are going to ruin everything the "makers" created. Alas, how quickly the one-percenters forget that many of them were former members of the 90% fraternity not too long ago! But, apparently, now that they have "arrived," they would like to keep the riffraff out.

As we wrap up things, it is useful to revisit the starting point of our endeavor in section 2.3: What is the purpose of a society? As we discussed, in the American context, it is to empower all individuals to realize life, liberty, and the pursuit of happiness in their own ways. Given this background, what is the purpose of a corporation situated on the American soil as an integral part of American society? And how is its purpose aligned with the purposes of the American people and their society?

The conventional view in economics and in management is that the purpose of a company is *only* to maximize profit and shareholder value. *But, is this purpose aligned with those of the people and their society?* We believe this alignment is vital for the long term. Ideally, they can all be aligned, but when profit maximization is carried out to extremes, through various measures that are harmful to the people and their society in the long run, the alignment begins to break down requiring new laws and regulations to restore fairness and stability.

Given the context of the foundational principles of our society, and its overall purpose, why should companies with abusive practices be allowed to exist among us? *What purpose do they serve in a society?* They do not exist in isolation, for they live and breathe among us, benefiting from our society's various and considerable resources. So, it is perfectly legitimate to ask what purpose they serve for our society. A society is a living and breathing organism, like the human body. When a cancerous growth, which exploits the body's resources for its own narrow benefits, begins to attack the health of the body, we know what to do. We remove it promptly. So, why should abusive companies, and their management, be treated any differently?

Again, it all boils down to the question of *telos*—the *purpose of life*, the *purpose of society*, and the *purpose of a corporation*. In the American context, the purpose of life is the pursuit of happiness, the purpose of the American society is to empower all citizens to realize this, and the purpose of any corporation on our soil *ought* to be to pursue its economic goals in alignment with these two purposes. At the very least, it should try not to get way out of line. We don't think the purpose of a corporation can be, or should be, sanitized of all meaningful elements of life and reduced to being a mere money spinning machine, disconnected from the purposes of life and society. Such a cynical approach might work for the short term but, as we have argued, is not sustainable in the long run. When a corporation loses sight of its fundamental and existential role, and thinks its only purpose is to maximize profits at all costs—even if it means gouging its customers for lifesaving medicines, for example—the society must act, and will act, justifiably, to correct such naked greed and flagrant abuse of market power. Such corporations are the ones that give the otherwise moral and productive free market capitalism a bad name.

If left unrecognized and uncorrected in midcourse, this form of myopic capitalism would eventually lead to violent structural adjustments as history teaches us. Adam Smith's vision of free-market capitalism is about maximizing stakeholder value—value enjoyed by employees, customers, investors, suppliers, and communities—not just shareholder value.

Even though our simple model does not account for these higher purposes explicitly, it nevertheless recognizes and accounts for the fundamental principle behind all this: fairness. As and when we move toward conscious capitalism, we need to figure out how to incorporate all this more completely in our models.

Our theory provides a theoretical justification, indeed a moral justification, for implementing the kind of corrective policies toward conscious capitalism recommended by leading economists, as well as by some business leaders, to empower the struggling majority. *This is simply a matter of restoring free-market fairness. Doing so is not Marxism. Doing so is not socialism. This is, in fact, the free-market capitalism that we all desire, the one envisioned by Adam Smith.* Our theory demonstrates that the ideal free-market capitalism could work for everyone, not just for the top 1%, if not gamed and subverted. This is the central message of our theory to the corporate boards, policy makers, politicians, and even the public at large. We should demand no less from our companies and governments!

While this may seem impossible in the current political climate in the United States (and perhaps elsewhere as well), we are hopeful, encouraged by the well-known quote "You can always count on the Americans to do the right thing, after they have tried everything else." This quote is often misattributed to Winston Churchill, but it seems to have originated from a variation of what Abba Eban (1967), the renowned Israeli statesman, said while visiting Japan in 1967. No matter who said it, there is a lot of truth to it, as our history demonstrates.

As Reich (2015, xvi) observes, "History provides some direction as well as some comfort, especially in America, which has periodically readapted the rules of the political economy to create a more inclusive society while restraining the political power of wealthy minorities at the top." He particularly highlights (Reich 2015, xvi) the Jacksonian movement in the 1830s to curb the power of the elites, antitrust laws in in early twentieth century, and the New Dealers limiting "the political power of large corporations and Wall Street, while enlarging the countervailing power of labor unions, small businesses, and small investors."

So let us conclude on that optimistic note that we have done the right things before, and we will do them again!

# Notes

## Preface

1. Venkatasubramanian et al. (2004).
2. Venkatasubramanian et al. (2006).
3. See section 2.2 pp. 31 for background on this name.
4. Venkatasubramanian (2007).
5. Venkatasubramanian (2009).
6. Venkatasubramanian (2010).
7. Venkatasubramanian et al. (2015).
8. Jaynes (1991a).
9. Pascal and Périer (1670).

## 1. Extreme Inequality in Income and Wealth

1. However, our theory considers only *wage income*, i.e., income earned from work, which is the dominant source of income for the majority of the population. We also use the term *pay* for wage in the book, as in CEO pay. Pay or wage refers to total compensation that includes base salary, bonus, options, etc.

2. Remember that our theory focuses only on *wage income*, i.e., income earned from work.

3. The phrase is commonly attributed to John F. Kennedy, who used it in a 1963 speech to combat criticisms that a dam project he was inaugurating was a pork barrel project. However, the phrase has been used more commonly to defend tax cuts and other policies where the initial beneficiaries are high-income earners. See Sorenson (2008).

## 2. Foundational Principles of a Fair Capitalist Society

1. The *veil of ignorance*, along with the *original position*, is a method of determining the morality of a certain issue based upon the following thought experiment. Parties to the original position know nothing about their ethnicity, gender, age, income, wealth, natural endowments, and position within the social order of society, etc. When such parties are selecting the principles for distribution of rights, positions, and resources in the society they will live in, this veil of ignorance prevents them from knowing about who they will be in that society. The idea is that parties subject to the veil of ignorance will make choices based upon moral considerations, since they will not be able to make choices based on self-interest or class interest (Wenar 2013).

2. According to luck egalitarianism, distributive justice should differentiate between results from one's bad luck, or those things over which one has no control, and those things that are the consequence of conscious options (Lamont and Favor 2014).

3. We leave it to the interested reader to explore on her own the important questions such as what "true and solid happiness" is, and whether one should pursue instead a meaningful life so that happiness *ensues*, as Viktor Frankl (1959) recommends.

## 3. Distributive Justice in a Hybrid Utopia

1. We introduced $Q$ to facilitate the reasoning. In general, one would not know the exact amount of future earnings. All we have, typically, is the hope that this job, and a series of ones after this, would lead us to a better life. So, the exact monetary value associated with better future prospects is generally unknown and unknowable. So the numerical value of $Q$ per se is not as pertinent as the general expectation of future benefits. Such an expectation is the basis for this component of the utility, the "future" utility, derived from the current job. In addition, while the benefits lie in the future, the cost is paid in the present, in the daily competition with one's peers, which necessitates the inclusion of the disutility term $-y \ln N_i$.

2. For an excellent, intuitive introduction to game theory and its applications, see Easley and Kleinberg (2010).

3. The prisoner's dilemma is a classic example in game theory that shows why two rational individuals might not cooperate, even if it appears that it is in their best interests to do so, and end up making suboptimal decisions. See Easley and Kleinberg (2010).

4. An excellent introduction to population games and evolutionary dynamics can be found in Sandholm (2010).

## 4. Statistical Thermodynamics and Equilibrium Distribution

1. Actually, what is inscribed is $S = k \log W$. Boltzmann himself did not write this particular form down with the constant $k$. This form was defined later by Max Planck in 1900.
2. As cited earlier, there are wonderful textbooks available for those who wish to dive deeply into this subject.
3. We don't write out the equations as they are not necessary for this discussion.

## 5. Fairness in Income Distribution

1. $S$ for entropy and $E$ for energy is standard notation in physics; hence we don't wish to deviate from that convention here. Since these occur only in the context of thermodynamics in our discussions, we believe confusion of these with salary and effort is really not a concern.
2. The need for this clarification was pointed out by an anonymous reviewer of the manuscript version of this book. I am grateful to the reviewer.
3. This previously unpublished result was obtained by V. Venkatasubramanian and Y. Luo and is presented here for the first time.
4. This previously unpublished result was obtained by V. Venkatasubramanian and Y. Luo and is presented here for the first time.
5. In our paper (Venkatasubramanian et al. 2015), following the result in Sandholm (2010), we observed that the equilibrium is *stable*. However, upon further reflection, we realized that it is indeed *asymptotically stable*, as proved above. This proof was derived by V. Venkatasubramanian and Y. Luo and is presented here for the first time.
6. One doesn't know what "small" means in real-life free-market perturbations. How much of a disturbance is small and how much is big in practice? This is one of the questions that needs to be investigated more carefully, as we mention in section 8.8 regarding future directions.
7. Similarly, there are important differences between *robustness* and *resilience*. All three terms—*stability, robustness,* and *resilience*—do have overlapping notions of failure avoidance, performance under failure, and recovery from failure, and hence tend to get used interchangeably. We are not getting into that discussion here as that would take us far from the main theme. Briefly, resilience comes from the word *resiliere*, which means to "bounce back." Suffice it to say that robustness is the ability to withstand disruption, whereas resilience is the ability to recover. A simple example would be a brick versus a block of sponge or rubber. A brick can withstand big hits, unyielding up to a point, beyond which it will break. A sponge will yield, deform, and adapt under

force, but will recover more or less to its original shape once the force is gone. Roughly speaking, robust systems focus on *resisting* failures while resilient systems focus on *recovering* from failures quickly. Robustness is the generalization of the notion of stability while resilience is the generalization of exponential stability.

8. As we pause to summarize our progress so far, it is useful to recognize that the concepts of *fairness* and *arbitrage* are intimately related through the actions of a free market. As we discussed in section 5.5.1, our notion of fairness is founded on equality and proportionality. We would protest treating two equal things unequally as unfair; treating two unequal things equally would also be deemed unfair. Regarding proportionality, consider this simple example. There are two workers, Bill and Bob, who work in the same company in similar, but slightly different, jobs. Bob's job requires contributing a little more than what Bill does in his job, and so Bob should be compensated more. However, if Bob were to be compensated a lot more than Bill, Bill would protest this disparity as unfair—"Why is he getting a lot more than me for doing just a little bit more?" This is seen as unfair because it violates the proportionality principle of fairness.

Arbitrage stems from pricing two things, which are equal or nearly equal, unequally, thereby providing a relatively easy profit-making opportunity. Consider a simple example. Jill works as a Python programmer making $50 per hour. It turns out a new programming language, Naga, has just emerged and is found to be very powerful for certain business applications. Therefore, there is a great demand for Naga programmers, at the market rate of $100 per hour. Jill finds out that it is not that hard for her to learn Naga. She realizes that this is a wonderful arbitrage opportunity that she can take advantage of—i.e., for a little more investment in effort and time, she can make a lot more money. So she acquires this skill in one month by taking online courses in the evenings. After a month, she quits her job and starts at a new one as Naga programmer making $100 per hour.

Now, as it turns out, Jill is not the only one who recognized this arbitrage opportunity. A lot of other Python programmers, who were also making $50 per hour, seize this opportunity and quickly acquire Naga programming skills with a couple of months of effort. This great increase in the supply of Naga programmers in short order begins to offset the demand, thereby reducing the compensation to $60 per hour. At the same time, the reduced supply of Python programmers, because of their switch to Naga jobs, increase their compensation from $50 per hour to $55 per hour. So, at steady state, when the market settles down, Python programmers make $55 per hour and Naga programmers make $60 per hour. That is, finally, one makes only a little more money for possessing a little more skill and contributing a little more. The arbitrage opportunity is gone now because the two nearly equal things have been priced properly. In fact, as we discussed in section 5.8.1, the salaries adjust in such a way that the effective utility of the Python programmers will be the same as that of the Naga programmers, at equilibrium in an ideal free-market. This is essentially what the ideal free-market delivers through its competitive pricing mechanism subject to supply and demand. The equality of effective utilities is equivalent to the equality of chemical potentials in statistical thermodynamics— i.e., the free-market's economic equilibrium is equivalent to chemical equilibrium in thermodynamics. And the lognormal distribution delivers this equality of effective utilities.

Thus, the free market's competitive mechanism eliminates arbitrage opportunities in effective utility by ensuring that no one derives a lot more utility for incurring just a little more cost—instead, a worker who makes a little more contribution makes only a little more money. This is the same as the proportionality principle of fairness discussed earlier. We, therefore, see that both fairness and arbitrage are founded on the same principles of equality and proportionality. This is the fundamental reason that an ideal free market can ensure the fairest outcome in income distribution. We believe this is an important insight about the free market that has not been recognized before. Our theory is the mathematical formulation of this insight.

9. In the context of social welfare, it is important to note the interesting connection between our equality of effective utilities result and the Nash bargaining solution. The Nash bargaining solution is obtained by maximizing the Nash product, $\pi_i(u_i - d_i)$, where $u_i$ is the utility and $d_i$ is the threat point of agent $i$, respectively. In our case, $h_i = u_i - d_i$, where $u_i$ is given by equation 3.3, and $d_i = v_i + w_i$, given by equations 3.2 and 3.4. Thus, the Nash product here is $\pi_i h_i$. Since we proved in section 5.8.2 that the total effective utility is $Nh^*$ is the social optimum, we can ask the following question: Given this total effective utility, when is the product of the individual utilities, i.e., the Nash product, maximized? The answer, of course, is when they are all equal. This well-known outcome is the result of the arithmetic mean–geometric mean relationship. Thus, the Nash bargaining solution gives us the same answer of equality of effective utilities, which we proved in two different ways—using the potential game and statistical mechanics approaches.

This connection also reveals a very important property of the Nash bargaining solution. Over the years, researchers have been somewhat puzzled by the Nash product. What does it really stand for? We can understand summing utilities, but why a product? While people understand it as a mathematical device, its economic content or purpose has been unclear. Our analysis reveals the essence of the Nash product. By maximizing a product of numbers that are constrained by their sum, what one is equalizing them. Equality of effective utilities is, of course, the fairest outcome as we have seen. Thus, by maximizing the Nash product one is really maximizing fairness subject to the Pareto efficiency constraint. This, I believe, is the true essence of the Nash product, very much like the true essence of game theoretic potential and entropy as we have seen.

## 6. Global Trends in Income Inequality

1. For this comparison, we are answering the following question: If any given country were to be an ideal society like BhuVai, with $S_{min}$ and $S_{max}$ as its cutoff salaries covering 99.9% of the population, what would be its income shares for the three tiers? These can be determined by computing the Lorenz curve using the equivalent standard normal distribution ($z$-table) that has the appropriate $\sigma$ given by 6.1.

Some have reported that an exponential distribution seems to fit the income distribution for the bottom ~95% using truncated datasets (see section 1.5.2). But, as we mentioned in 1.5.2, truncated datasets can quite easily lead to misidentifications. A truncated lognormal distribution, covering the range from its mode to its tail on the right, may be approximated as an exponential distribution. This is particularly true for income data, for they are typically not terribly precise. Furthermore, the $u - v$ part of effective utility function given by equation 3.6, for the parameter values given in table 6.2 (which are representative of the 12 countries reported in section 6.2), does not have a pronounced curvature, particularly for the region corresponding to the truncated part of the lognormal distribution. Therefore, this may be approximated as a linear function, which would then result in an exponential distribution after entropy maximization, explaining why some have reported the exponential fit. I believe that this is only an approximation, may even be a good approximation, but the true underlying distribution is lognormal.

2. I am grateful to V. Govind Manian for pointing this out to me.

## 7. What Is Fair Pay for Executives?

1. This analysis was first published in my 2009 paper (Venkatasubramanian 2009), hence the use of the 2008 data. The main conclusions are still valid as the trends concerning CEO pay have not changed much between 2008 and 2016.

## 8. Final Synthesis and Future Directions

1. One could argue whether the adjective *maximum* is really needed here. Doesn't fairness as a design principle capture the intent? Isn't something either fair or unfair? Do we really need to say something is maximally fair? Isn't this

like saying someone is maximally pregnant? One is either pregnant or not pregnant. Well, we use the qualifier to capture the degree of closeness or balance in a distribution. For example, a 51:49 split is a lot closer to the fair allocation of 50:50 than a 90:10 split. Thus, while a 50:50 share is the definition of fairness, we use the qualifier *maximum* to emphasize the degree of proximity to the fair allocation.

2. I would like to note an interesting aside here. When I first described the central ideas of this theory in a manuscript submitted to the journal *Entropy* in 2009, I was unaware of the contributions of Rawls, Nozick, and others in political philosophy. Their contributions were pointed out to me by an alert reviewer in a critical review of my manuscript. I am most grateful to this person for the review helped me develop my theory in a more comprehensive manner, going all the way from molecules to humans, from statistical mechanics and system dynamics to political philosophy. So, in my revised submission, I broadened my framework to compare and contrast statistical teleodynamics with the philosophical theories, along the lines discussed above.

Though my theory was inspired by concepts from game theory, statistical thermodynamics, and information theory, I found it (and I still do) rather remarkable that the resultant statistical teleodynamic framework shares such a close resemblance with the structure and elements of the theories of Rawls and Nozick (and Dworkin). In the context of economics (and maybe in the broader context of sociology as well), I believe that the statistical teleodynamic framework naturally combines the central ideas of Rawls and Nozick (and Dworkin), even though its development was not motivated by this objective (in fact, as noted, I wasn't even aware of their work). In addition, our theory goes one step further by making quantitative predictions which can be compared with reality. None of the philosophical theories accomplish this.

3. We are aware of exceptions such as broken symmetries and charge-parity symmetry (CP-Symmetry) violation, but discussing those here would take us far from the main discussion.

4. This question was indeed raised by an anonymous reviewer of the manuscript version of this book. I am grateful to the reviewer for highlighting this concern, for it has given me an opportunity to explain why I think my theory could work.

5. This result has the classic setup of a good news–bad news joke:

Me: Hi! I have got both good news and bad news for you!

Free-market fundamentalist (looking up from his Bloomberg Terminal, annoyed): Oh, it's you! You are always trouble! Anyway, what's the good news?

Me: Well, I can prove mathematically that the ideal free-market society is fair, morally justified, socially optimal, and stable!

FMF: I knew it! I knew it! That's fantastic! What's the bad news?

Me: The society looks like Scandinavia!

FMF: I knew you were trouble!

# Bibliography

Ackerman, F. 2002. "Still Dead After All These Years: Interpreting the Failure of General Equilibrium Theory." *Journal of Economic Methodology* 9(2): 119–39.

Adidevananda, S., trans. 2009. *Sri Ramanuja Gita Bhasya*. Mylapore, Madras: Sri Ramakrishna Math.

Ahituv, A., and R. I. Lerman. 2007. "How Do Marital Status, Work Effort, and Wage Rates Interact?" *Demography* 44(3): 623–47.

Akerlof, G. A., and R. J. Shiller. 2009. *Animal Spirits: How Human Psychology Drives the Economy, and Why It Matters for Global Capitalism*. Princeton, NJ: Princeton University Press.

Akerlof, G. A., and J. L. Yellen. 1986. *Efficiency Wage Models of the Labor Market*. London: Cambridge University Press.

Alvaredo, F., A. B. Atkinson, T. Piketty, and E. Saez. 2015. World Top Incomes Database (Top 10%, Top 1% Income Shares, p99.9 Threshold, Average Income per Tax Unit, Bottom 90% Average Income, and Top 0.5% Average Income; Australia, Canada, Denmark, France, Germany, Japan, Netherlands, Norway, Sweden, Switzerland, United Kingdom, and United States). http://wid.world/.

Anderson, S., J. Cavanagh, C. Collins, M. Lapham, and S. Pizzigati. 2008. *Executive Excess 2008: How Average Taxpayers Subsidize Runaway Pay*. Washington, DC: Institute for Policy Studies, http://www.ips-dc.org/executive_excess_2008_how_average_taxpayers_subsidize_runaway_pay/.

Aström, K. J., and R. M. Murray. 2008. *Feedback Systems: An Introduction for Scientists and Engineers*. Princeton, NJ: Princeton University Press.

Atkinson, A. B. 2015. *Inequality: What Can Be Done?* Cambridge, MA: Harvard University Press.

Axtell, R. 2014. "Endogenous Dynamics of Firms and Labor with Large Numbers of Simple Agents." http://www.lem.sssup.it/WPLem/documents/axtell.pdf.

Ball, R., and P. Holmes. 2008. "Dynamical Systems, Stability, and Chaos." *Frontiers in Turbulence and Coherent Structures* 6:1–27.

Bares, A. 2013. "CEO Pay: Why Not More Outrage over Other Highly Paid People?" ERE Media. http://www.eremedia.com/tlnt/ceo-pay-why-not-more-outrage-over-other-highly-paid-people/.

Bassett, D., and F. Claveau. 2011. "The Economic Entomologist: An Interview with Alan Kirman." *Erasmus Journal for Philosophy and Economics* 4(2): 42–66.

Baumgartner, S. 2005. "Thermodynamic Models: Rationale, Concepts, and Caveats." In *Modelling in Ecological Economics*, ed. J. L. Proops and P. Safonov. Northhampton, MA: Edward Elgar.

Bausor, R. 1987. "Liapounov Techniques in Economic Dynamics and Classical Thermodynamics." Paper presented at the meeting of the American Economic Association, Chicago.

Bebchuk, L. A., and J. M. Fried. 2005. "Pay Without Performance: Overview of the Issues." *Journal of Applied Corporate Finance* 17(4): 8–22.

———. 2006. *Pay Without Performance: The Unfulfilled Promise of Executive Compensation*. Cambridge, MA: Harvard University Press.

Beckmann, D. M., and A. R. Simon. 1965. *Grace at the Table: Ending Hunger in God's World*. Mahwah, NJ: Paulist Press.

Benabou, R., and E. A. Ok. 2001. *Mobility as Progressivity: Ranking Income Processes according to Equality of Opportunity*. Technical report. Cambridge, MA: National Bureau of Economic Research.

Bentham, J. 1982. *Collected Works of Bentham, An introduction to the Principles of Morals and Legislation*. Oxford: Clarendon Press.

Bernoulli, J. 1713. *Ars Conjectandi*. Impensis Thurnisiorum, fratrum.

Bloxham, E. 2011. "How Can We Address Excessive CEO Pay?" *Fortune*, April 13.

Blume, L. E. 1993. "The Statistical Mechanics of Strategic Interaction." *Games and Economic Behavior* 5(3): 387–424.

Bodhi, B. 2005. *In the Buddha's Words: An Anthology of Discourses from the Pali Canon*. Somerville, MA: Wisdom.

Böhringer, C., and A. Löschel. 2013. *Empirical Modeling of the Economy and the Environment*, vol. 20. Berlin: Springer Science & Business Media.

Bolton, G. E., and A. Ockenfels. 2000. "Erc: A Theory of Equity, Reciprocity, and Competition." *American Economic Review* 90(1): 166–93.

Boltzmann, L. 1964. *Lectures on Gas Theory*. Berkeley: University of California Press.

Bookstaber, R., P. Glasserman, G. Iyengar, Y. Luo, V. Venkatasubramanian, and Z. Zhang. 2015. "Process Systems Engineering as a Modeling Paradigm for Analyzing Systemic Risk in Financial Networks." *Journal of Investing* 24(2): 147–62.

Bouchaud, J.P., and M. Mézard. 2000. "Wealth Condensation in a Simple Model of Economy." *Physica A: Statistical Mechanics and Its Applications* 282(3): 536–45.

Box, G. E. P., and N. R. Draper. 1987. *Empirical Model Building and Response Surface*. New York: Wiley.

Brown, V. 1994. *Adam Smith's Discourse: Canonicity, Commerce, and Conscience*. East Sussex, UK: Psychology Press.

Buffett, W. 2004. Letter to Shareholders of Berkshire Hathaway Inc., February.

Buren, P. V. 2014. "In Today's America, a Rising Tide Lifts All Yachts." *Nation*, June 3. http://www.thenation.com/article/todays-america-rising-tide-lifts-all-yachts/.

Byers, N. 1998. "E. Noether's Discovery of the Deep Connection Between Symmetries and Conservation Laws." arXiv preprint, https://arxiv.org/abs/physics/9807044.

Cadenillas, A., J. Cvitanic, and F. Zapatero. 2002. "Executive Stock Options with Effort Disutility and Choice of Volatility." http://people.hss.caltech.edu/~cvitanic/PAPERS/exec.pdf.

Cassidy, J. 2008. "The Minsky Moment." *New Yorker*, Feb 4.

—. 2014a. "Forces of Divergence." *New Yorker*, March 31.

—. 2014b. "Piketty's Inequality Story in Six Charts." *New Yorker*, March 26.

Chakrabarti, B. K., A. Chakraborti, S. R. Chakravarty, and A. Chatterjee. 2013. *Econophysics of Income and Wealth Distributions*. London: Cambridge University Press.

Champernowne, D. G. 1953. "A Model of Income Distribution." *Economic Journal* 63(250): 318–51.

Champernowne, D. G., and F. A. Cowell. 1998. *Economic Inequality and Income Distribution*. London: Cambridge University Press.

Charles, K. K., and E. Hurst. 2003. "Intergenerational Wealth Correlations." *Journal of Political Economy* 111(6): 1155–82.

Chatterjee, A., S. Sinha, and B. K. Chakrabarti. 2007. "Economic Inequality: Is It Natural?" *Current Science* 92(10): 1383–89.

Chatterjee, A., S. Yarlagadda, and B. K. Chakrabarti. 2005. *Econophysics of Wealth Distributions: Econophys-Kolkata I*. Berlin: Springer Science and Business Media.

Cho, A. 2014. "Physicists Say It's Simple." *Science* 344(6186): 828.

Clausius, R. 1867. *The Mechanical Theory of Heat: With Its Applications to the Steam-Engine and to the Physical Properties of Bodies*. London: J. van Voorst.

Cohen, G. A. 2008. *Rescuing Justice and Equality*. Cambridge, MA: Harvard University Press.

Crystal, G. S. 1991. *In Search of Excess: The Overcompensation of American Executives*. New York: Norton.

Darwin, C. 1859. *The Origin of Species*. London: John Murray.

de Laplace, P. S. 1814. "Essai philosophique sur les probabilités." *Academie des Sciences, Oeuvres Complètes de Laplace* 7.

DeLuca, D., and S. Vivekananda. 2006. *Pathways to Joy: The Master Vivekananda on the Four Yoga Paths to God*. Novato, CA: New World Library.

Domhoff, G. W. 2006. "Power in America: Wealth, Income, and Power." *Who Rules America?* (blog), http://www2.ucsc.edu/whorulesamerica/power/wealth.html.

Dossani, R. 2002. "Telecommunications Reform in India." *India Review* 1(2): 61–90.

Dragulescu, A., and V. M. Yakovenko. 2000. "Statistical Mechanics of Money." *The European Physical Journal B-Condensed Matter and Complex Systems* 17(4): 723–29.

Drucker, P., and B. Schlender. 2004. "Peter Drucker Sets Us Straight: The 94-Year-Old Guru Says That Most People Are Thinking All Wrong About Jobs, Debt, Globalization, and Recession." *Fortune Magazine*.

Dworkin, R. 1981a. "What Is Equality? Part 1: Equality of Welfare." *Philosophy & Public Affairs* 10(3): 185–246.

——. 1981b. "What Is Equality? Part 2: Equality of Resources." *Philosophy & Public Affairs* 10(4): 283–345.

——. 2001. *Sovereign Virtue: The Theory and Practice of Equality*. Cambridge, MA: Harvard University Press.

Easley, D., and J. Kleinberg. 2010. *Networks, Crowds, and Markets: Reasoning About a Highly Connected World*. London: Cambridge University Press.

Eban, A. 1967. http://quoteinvestigator.com/2012/11/11/exhaust-alternatives/, September 14.

Ebeling, Richard M. 1999. "Friedrich A. Hayek: A Centenary Appreciation." *Freeman* 49(5), 31.

Einstein, A. 2010. *The Ultimate Quotable Einstein*. Princeton, NJ: Princeton University Press.

Etter, L. 2006. "Hot Topic: Are CEOs Worth Their Weight in Gold?" *Wall Street Journal*, January 21, A7.

Farmer, J. D., E. Smith, and M. Shubik. 2005. "Economics: The Next Physical Science?" *Physics Today* 58(9): 37–42.

Fehr, E., E. Kirchler, A. Weichbold, and S. Gächter. 1998. "When Social Norms Overpower Competition: Gift Exchange in Experimental Labor Markets." *Journal of Labor Economics* 16(2): 324–51.

Fehr, E., and K. M. Schmidt. 1999. "A Theory of Fairness, Competition, and Cooperation." *Quarterly Journal of Economics* 114: 817–68.

Feldman, F., and B. Skow. 2015. "Desert." In *Stanford Encyclopedia of Philosophy*. https://plato.stanford.edu/entries/desert/.

Fernando, V. 2009. "CEOs Don't Make Jack Compared to Athletes." *Business Insider*, http://www.businessinsider.com/sports-stars-are-way-more-overpaid-than-ceos-2009-11.

Feynman, R. P. 1965. *The Character of Physical Law*. New York: Random House.

Fisher, F. M. 1989. *Disequilibrium Foundations of Equilibrium Economics*. Number 6. London: Cambridge University Press.

—. 2011. "The Stability of General Equilibrium: What Do We Know and Why Is It Important?" In *General Equilibrium Analysis: A Century after Walras*, ed. P. Bridel, 34–45. London: Routledge, Taylor and Francis.

Fleischacker, S. 2004a. *On Adam Smith's Wealth of Nations*. Princeton, NJ: Princeton University Press.

—. 2004b. *A Short History of Distributive Justice*. Cambridge, MA: Harvard University Press.

Foley, D. K. 1994. "A Statistical Equilibrium Theory of Markets." *Journal of Economic Theory* 62(2): 321–45.

—. 1996. "Statistical Equilibrium in a Simple Labor Market." *Metroeconomica* 47(2): 125–47.

—. 1999. "Statistical Equilibrium in Economics: Method, Interpretation, and an Example." Notes for the ISER Summer School in Siena, July 4–11.

—. 2010. "What's Wrong with the Fundamental Existence and Welfare Theorems?" *Journal of Economic Behavior & Organization* 75(2): 115–31.

Frankl, V. E. 1959. *Man's Search for Meaning*. Boston: Beacon.

Freeland, C. 2011. "The Lottery Mentality." *New York Times*, March 21.

Gallegati, M., S. Keen, T. Lux, and P. Ormerod. 2006. "Worrying Trends in Econophysics." *Physica A: Statistical Mechanics and Its Applications* 370(1): 1–6.

Garofalo, P. 2013. "What We Can Learn from Switzerland's CEO Pay Cap Vote." *U.S. News and World Report*, November 25. http://www.usnews.com/opinion/blogs/pat-garofalo/2013/11/25/the-importance-of-switzerlands-112-ceo-pay-cap-vote.

Gaus, G. 2010. *The Order of Public Reason: A Theory of Freedom and Morality in a Diverse and Bounded World*. London: Cambridge University Press.

Georgescu, P. 2015. "Capitalists, Arise: We Need to Deal with Income Inequality." *New York Times*, August 7.

Georgescu-Roegen, N. 1987. "The Entropy Law and the Economic Process in Retrospect." *Schriftenreihe des IÖW* 5: 87.

Gibbs, J. W. 1906. *The Scientific Papers of J. Willard Gibbs*. New York: Longmans, Green and Company.

—. 1928. *The Collected Works of J. Willard Gibbs*, vol. 1. New Haven, CT: Yale University Press.

Gini, C. 1921. "Measurement of Inequality of Incomes." *Economic Journal* 31(121): 124–26.

Güth, W. 1995. "On Ultimatum Bargaining Experiments—A Personal Review." *Journal of Economic Behavior & Organization* 27(3): 329–44.

Hallock, K. F. 2012a. *Pay: Why People Earn What They Earn and What You Can Do Now to Make More.* London: Cambridge University Press.

——. 2012b. "Top Athlete Pay." Technical report. Institute for Compensation Studies, Cornell University ILR School.

Hamilton, C. V. 2008. "Why Did Jefferson Change 'Property' to the 'Pursuit of Happiness'?" *History News Network*, http://historynewsnetwork.org/article/46460.

Hargreaves, D. 2014. "Can We Close the Pay Gap?" *New York Times*, March 29. http://opinionator.blogs.nytimes.com/2014/03/29/can-we-close-the-pay-gap/?_r=0.

Hayek, Friedrich von. 1944. *The Road to Serfdom.* London: Routledge.

Heilbroner, R. L. 1999. *The Worldly Philosophers.* New York: Simon and Schuster.

Helbing, D., and A. Kirman. 2013. "Rethinking Economics Using Complexity Theory." *Real-World Economics Review*, 64. https://rwer.wordpress.com/2013/07/02/rwer-issue-64/.

Herbert, C. B. 1985. *Thermodynamics and an Introduction to Thermostatistics.* New York: Wiley.

Hofbauer, J., and K. Sigmund. 1998. *Evolutionary Games and Population Dynamics.* London: Cambridge University Press.

Hogg, R. V., and A. T. Craig. 1995. *Introduction to Mathematical Statistics*, 5th ed. London: Macmillan.

Holmberg, S., and M. Schmitt. 2014. "The Overpaid CEO." *Democracy* no. 34, p. 60.

Holmstrom, B., and P. Milgrom. 1991. "Multitask Principal-Agent Analyses: Incentive Contracts, Asset Ownership, and Job Design." *Journal of Law, Economics, & Organization* 7(special issue): 24–52.

Hozo, S. P., B. Djulbegovic, and I. Hozo. 2005. "Estimating the Mean and Variance from the Median, Range, and the Size of a Sample." *BMC Medical Research Methodology* 5(1): 13.

Jaynes, E. T. 1957a. "Information Theory and Statistical Mechanics." *Physical Review* 106(4): 620.

——. 1957b. "Information Theory and Statistical Mechanics. II." *Physical Review* 108(2): 171.

——. 1979. "Where Do We Stand on Maximum Entropy?" *The Maximum Entropy Formalism.* Cambridge, MA: MIT Press, 15–118.

——. 1985. "Where Do We Go from Here?" In *Maximum-Entropy and Bayesian Methods in Inverse Problems*, ed. C. R. Smith and W. Grandy, 21–58. New York: Springer.

———. 1991a. "How Should We Use Entropy in Economics?" Technical report. Working Paper. http://bayes.wustl.edu/etj/articles/entropy.in.economics.pdf.

———. 1991b. "Notes on Present Status and Future Prospects." In *Maximum Entropy and Bayesian Methods*, ed. W. T. Grady Jr. and L. H. Schick, 1–13. New York: Springer.

Jen, E. 2005. "Stable or Robust? What's the Difference?" In *Robust Design*, ed. E. Jen, chap. 1. Oxford: Oxford University Press.

Johnson, S. 1787. *The History of Rasselas, Prince of Abissinia: A Tale*, vol. 4. Harrison House.

Jones, K. 2009a. "C.E.O. Compensation: The Pay at the Top." *New York Times*.

———. 2009b. "Who Moved My Bonus? Executive Pay Makes a U-Turn." *New York Times*, April 4. http://www.nytimes.com/2009/04/05/business/05comp.html.

Jorgenson, D. W. 1960. "A Dual Stability Theorem." *Econometrica: Journal of the Econometric Society* 28(4): 892–99.

Kaplow, L., and S. Shavell. 2001. "Fairness versus Welfare." *Harvard Law Review* 114(4): 961–1388.

Kapur, J. N. 1989. *Maximum-Entropy Models in Science and Engineering*. New York: Wiley.

Kato, T., and K. Kubo. 2006. "CEO Compensation and Firm Performance in Japan: Evidence from New Panel Data on Individual CEO Pay." *Journal of the Japanese and International Economies* 20(1): 1–19.

Katz, L. F. 1986. "Efficiency Wage Theories: A Partial Evaluation." In *NBER Macroeconomics Annual 1986*, vol. 1, 235–90. Cambridge, MA: MIT Press.

Keen, S. 2011. *Debunking Economics: Revised and Expanded Edition: The Naked Emperor Dethroned?* New York: Zed.

Keister, L. A. 2005. *Getting Rich: America's New Rich and How They Got That Way*. London: Cambridge University Press.

Kelvin, L. 1883. "The Six Gateways of Knowledge." http://zapatopi.net/kelvin/papers/the_six_gateways_of_knowledge.html.

Kent, S. 2016. "BP Shareholders Reject Oil Giant's Pay Policy." *Wall Street Journal*, April 14.

Kenworthy, L. 2015. *The Good Society*. Unpublished.

Keynes, J. M. 1921. *A Treatise on Probability*. Vol. 8 of *The Collected Writings of John Maynard Keynes*. Cambridge: Cambridge University Press.

———. 1935. *General Theory of Employment, Interest and Money*. New Delhi, India: Atlantic Publishers & Distributors.

———. 1963. *Essays in Persuasion*. New York: Norton.

Khinchin, A. I. 1957. *Mathematical Foundations of Information Theory*, vol. 434. North Chelmsford, MA: Courier.

Kiatpongsan, S., and M. I. Norton. 2014. "How Much (More) Should CEOs Make? A Universal Desire for More Equal Pay." *Perspectives on Psychological Science* 9(6): 587–93.

Kikkawa, M. 2009. "Statistical Mechanics of Games—Evolutionary Game Theory." *Progress of Theoretical Physics Supplement* 179: 216–26.

Kirman, A. 1989. "The Intrinsic Limits of Modern Economic Theory: The Emperor Has No Clothes." *Economic Journal* 99(395): 126–39.

———. 2010. "The Economic Crisis Is a Crisis for Economic Theory." *CESifo Economic Studies* 56(4): 498–535.

Kirman, A. 2011. "Walras's Unfortunate Legacy." In *General Equilibrium Analysis: A Century after Walras*, ed. P. Bridel, 109–133. London: Routledge, Taylor and Francis.

Kirman, A., et al. 2006. "Demand Theory and General Equilibrium: From Explanation to Introspection, a Journey down the Wrong Road." *History of Political Economy* 38: 246.

Klinger, S., S. Pizzigati, and S. Anderson. 2013. "Executive Excess 2013: Bailed Out, Booted, and Busted." Institute for Policy Studies, http://www.ips-dc.org/executive-excess-2013/.

Kodjak, A. 2016. "A Peek inside Turing Pharmaceuticals: 'Another $7.2 Million. Pow!'" *NPR*, February 2. http://www.npr.org/sections/health-shots/2016/02/02/465284148/-another-7-2-million-pow-a-peek-inside-turing-pharmaceuticals.

Koppelman, A. 2012. "Review of Free Market Fairness by John Tomasi." *Notre Dame Philosophical Reviews* 2012.05.05. http://ndpr.nd.edu/news/30638-free-market-fairness/.

Kristof, N. 2014. "We're Not No. 1! We're Not No. 1!" *New York Times*, April 3.

Krugman, P. 2010. "Economics Is Not a Morality Play." *New York Times*, September 28. http://krugman.blogs.nytimes.com/2010/09/28/economics-is-not-a-morality-play/.

———. 2014. "Inequality Is a Drag." *New York Times*, August 8. http://www.nytimes.com/2014/08/08/opinion/paul-krugman-inequality-is-a-drag.html online.

———. 2015. "Twin Peaks Planet." *New York Times*, January 2. http://www.nytimes.com/2015/01/02/opinion/paul-krugman-twin-npeaks-planet.html online.

———. 2016. "Trade, Labor, and Politics." *New York Times*, March 28. http://www.nytimes.com/2016/03/28/opinion/trade-labor-and-politics.html online.

Laffont, J.-J., and J. Tirole. 1987. "Auctioning Incentive Contracts." *Journal of Political Economy* 95: 921–37.

———. 1988. "The Dynamics of Incentive Contracts." *Econometrica: Journal of the Econometric Society* 56: 1153–75.

———. 1993. *A Theory of Incentives in Procurement and Regulation.* Cambridge, MA: MIT Press.

Lambert, F. L. 2002. "Disorder—A Cracked Crutch for Supporting Entropy Discussions." *Journal of Chemical Education* 79(2): 187.

Lamont, J., and C. Favor. 2014. "Distributive Justice." In *Stanford Encyclopedia of Philosophy.* https://plato.stanford.edu/entries/justice-distributive/.

Lan, T., D. Kao, M. Chiang, and A. Sabharwal. 2010. *An Axiomatic Theory of Fairness in Network Resource Allocation.* arXiv preprint, https://arxiv.org/abs/cs/0906.0557.

Lazear, E. P., and K. L. Shaw. 2007. *Personnel Economics: The Economist's View of Human Resources.* Technical report. Cambridge, MA: National Bureau of Economic Research.

Lebacqz, K. 1986. *Six Theories of Justice.* Minneapolis: Augsburg.

Lederman, L. M., and C. T. Hill. 2004. *Symmetry and the Beautiful Universe.* Amherst, NY: Prometheus Books.

Levy, M., and S. Solomon. 1996. "Power Laws Are Logarithmic Boltzmann Laws." *International Journal of Modern Physics C* 7(4): 595–601. C. S.

Lewis, *Mere Christianity*, New York, Simon & Schuster Touchstone, 1996.

Lincoln, A. 1863. Draft of the Gettysburg Address: Nicolay copy, November 1863; series 3, general correspondence, 1837–1897. *The Abraham Lincoln Papers at the Library of Congress, Manuscript Division*, Washington: American Memory Project, 2000-02.

Locke, J. 1689. *The Works of John Locke,* vol. 1. London: Bell and Daldy.

Lublin, J. 2016. "Companies Wind Up in the 'Penalty Box' on Executive Pay." *Wall Street Journal*, April 20.

Lucas, R. E. 2002. "The Industrial Revolution: Past and Future." In *Lectures on Economic Growth*, ed. Robert E. Lucas, Jr. Cambridge, MA: Harvard University Press, 109–188.

Mackey, J., and R. Sisodia. 2014. *Conscious Capitalism.* Boston: Harvard Business Review Press.

Maharshi, R. 2000. *Talks with Ramana Maharshi: On Realizing Abiding Peace and Happiness.* Carlsbad, CA: Inner Directions.

Maremont, M. 2016. "Epipen Maker Dispenses Outsize Pay." *Wall Street Journal*, September 13.

Milanovic, B. 2002. "True World Income Distribution, 1988 and 1993: First Calculation Based on Household Surveys Alone." *Economic Journal* 112(476): 51–92.

———. 2012. *The Haves and the Have-Nots: A Brief and Idiosyncratic History of Global Inequality.* New York: Basic Books.

Mill, J. S. 1957. *Utilitarianism.* New York: Bobbs-Merrill.

Minsky, H. P., and H. Kaufman. 1986. *Stabilizing an Unstable Economy.* New York: McGraw-Hill.

Mirowski, P. 1989. *More Heat Than Light: Economics as Social Physics, Physics as Nature's Economics*. London: Cambridge University Press.

Mishel, L., J. Bivens, E. Gould, and H. Shierholz. 2012. *The State of Working America*. Ithaca, NY: Cornell University Press.

Mishel, L., and A. Davis. 2014. *CEO Pay Continues to Rise as Typical Workers Are Paid Less*. Washington, D.C.: Economic Policy Institute.

Mitzenmacher, M. 2004. "A Brief History of Generative Models for Power Law and Lognormal Distributions." *Internet Mathematics* 1(2): 226–51.

Mohammed, A. 2015. "Deepening Income Inequality." *World Economic Forum, Outlook on the Global Agenda*, http://reports.weforum.org/outlook-global-agenda-2015/top-10-trends-of-2015/1-deepening-income-inequality/.

Monderer, D., and L. S. Shapley. 1996. "Potential Games." *Games and Economic Behavior* 14(1): 124–43.

Montroll, E. W., and M. F. Shlesinger. 1982. "On $1/f$ Noise and Other Distributions with Long Tails." *Proceedings of the National Academy of Sciences* 79(10): 3380–83.

Mood, A. M., F. A. Graybill, and D. C. Boes. 1974. *Introduction to the Theory of Statistics*, 3rd ed. New York: McGraw-Hill.

Nalebuff, B. J., and J. E. Stiglitz. 1983. "Prizes and Incentives: Towards a General Theory of Compensation and Competition." *Bell Journal of Economics* 4(1): 21–43.

Nash, J. 1951. "Non-cooperative Games." *Annals of Mathematics* 54(2): 286–95.

Nash, J. F. 1950. "Equilibrium Points in $n$-Person Games." *Proceedings of the National Academy of Sciences USA* 36(1): 48–49.

Nash, L. K. 1970. *Elements of Classical and Statistical Thermodynamics*, vol. 5238. Reading, MA: Addison-Wesley.

———. 2012. *Elements of Statistical Thermodynamics*. North Chelmsford, MA: Courier.

Nocera, J. 2014. "C.E.O. Pay Goes Up, Up and Away!" *New York Times*, April 15.

Norton, M. I., and D. Ariely 2011. "Building a Better America—One Wealth Quintile at a Time." *Perspectives on Psychological Science* 6(1), 9–12.

Nowak, M. A., K. M. Page, and K. Sigmund. 2000. "Fairness versus Reason in the Ultimatum Game." *Science* 289(5485): 1773–75.

Nozick, R. 1974. *Anarchy, State, and Utopia*. New York: Basic Books.

Nussbaum, M. C. 2011. *Creating Capabilities: The Human Development Approach*. Cambridge, MA: Harvard University Press.

OECD. 2015. "Income Distribution and Poverty: By Country—Inequality." http://stats.oecd.org/index.aspx?queryid=46189.

O'Neill, M., and T. Williamson. 2012. "Free Market Fairness: Is There a Moral Case for Free Markets?" *Boston Review*, Nov. 5. http://bostonreview.net/books-ideas/free-market-fairness.

Ormerod, P. 2010. "The Current Crisis and the Culpability of Macroeconomic Theory." *Twenty-First Century Society* 5(1): 5–18.

Pandit, M. P. 1974. *Mystic Approach to the Veda & the Upanishad*. India: Ganesh & Co.

Pareto, V. 1897. *Cours d'économie politique*. Paris: Droz.

Pascal, B., and E. Périer. 1670. *Pensées de Mr. Pascal sur la religion et sur quelques autres sujets, qui ont esté trouvées après sa mort parmy ses papiers*, vol. 1. Chez Abraham Wolfganck suivant la copie imprimée à Paris.

Perline, R. 2005. "Strong, Weak and False Inverse Power Laws." *Statistical Science* 20(1): 68–88.

Piketty, T., and A. Goldhammer. 2014. *Capital in the Twenty-First Century*. Cambridge, MA: Belknap Press.

Piketty, T. and E. Saez. 2003. "Income Inequality in the United States, 1913–1998." *Quarterly Journal of Economics* 118(1): 1–41.

Popken, B. 2016. "Mylan CEO's Pay Rose over 600 Percent as Epipen Price Rose 400 Percent." *NBC News*, August 23. http://www.nbcnews.com/business/consumer/mylan-execs-gave-themselves-raises-they-hiked-epipen-prices-n636591.

Popkin, R. H. 1966. *The Philosophy of the 16th and 17th Centuries*. New York: Simon and Schuster.

Proops, J. L., and P. Safonov. 2004. *Modelling in Ecological Economics*. Northhampton, MA: Edward Elgar Publishing.

Rakoff, J. S. 2014. "The Financial Crisis: Why Have No High-Level Executives Been Prosecuted?" *New York Review of Books*, January 9.

Rawls, J. 1971. *A Theory of Justice*. Cambridge, MA: Harvard University Press.

———. 2001. *Justice as Fairness: A Restatement*. Cambridge, MA: Harvard University Press.

Reich, R. B. 2015. *Saving Capitalism: For the Many, Not the Few*. New York: Knopf.

Reid, C. 1996. *Hilbert*. Gottingen: Copernicus.

Reif, F. 1965. *Fundamentals of Statistical and Thermal Physics*. New York: McGraw-Hill.

Richmond, P., S. Hutzler, R. Coelho, and P. Repetowicz. 2006. *A Review of Empirical Studies and Models of Income Distributions in Society*. Berlin: Wiley-VCH.

Richmond, P., J. Mimkes, and S. Hutzler. 2013. *Econophysics and Physical Economics*. New York: Oxford University Press.

Roemer, J. E. 1996. *Theories of Distributive Justice*. Cambridge, MA: Harvard University Press.

Rogers, B. 2000. "Dworkin's Desert Island." *Prospect: The Leading Magazine of Ideas*, August 20.

Roosevelt, F. D. 1937. "Presidential Inaugural Address, January 20, 1937." http://historymatters.gmu.edu/d/5105/.

Roosevelt, T. 1906. "State of the Union Address, December 3, 1906." http://teachingamericanhistory.org/library/document/state-of-the-union-address-part-i-11/.

Rosenthal, R. W. 1973. "A Class of Games Possessing Pure-Strategy Nash Equilibria." *International Journal of Game Theory* 2(1): 65–67.

Saez, E., and G. Zucman. 2014. *Wealth Inequality in the United States Since 1913: Evidence from Capitalized Income Tax Data*. Technical report. Cambridge, MA: National Bureau of Economic Research.

Salaman, E. 1955. "A Talk with Einstein." *Listener* 54: 370–71.

Samuelson, P. A. 1972. Maximum Principles in Analytical Economics." *American Economic Review* 62: 249–62.

——. 1990. "Gibbs in Economics." *Proceedings of the Gibbs Symposium*, Providence, RI: American Institute of Physics, 255–67.

Sandel, M. J. 2009. *Justice: What's the Right Thing to Do?* New York: Farrar, Straus, and Giroux.

Sandholm, W. H. 2010. *Population Games and Evolutionary Dynamics*. Cambridge, MA: MIT Press.

Scanlon, T. 2005. *When Does Equality Matter?* Typescript. Cambridge, MA: Harvard University.

Schmidtz, D. 2006. *The Elements of Justice*. London: Cambridge University Press.

Schoenberg, E. 2011. "Zombie Economics and Just Deserts: Why the Right Is Winning the Economic Debate." *Huffington Post*, January 27 (updated May 25). http://www.huffingtonpost.com/eric-schoenberg/zombie-economics-and-just-b_806110.html.

Sen, A., and J. E. Foster. 1997. *On Economic Inequality*. London: Oxford University Press.

Shankar, R. 2014. *Fundamentals of Physics: Mechanics, Relativity, and Thermodynamics*. New Haven, CT: Yale University Press.

Shannon, C. E. 1948. A Mathematical Theory of Communication, *Bell System Technical Journal* 27: 379–423, 623–656.

Simpson, J. A., E. S. Weiner, et al. 1989. *Oxford English Dictionary*, vol. 2. Oxford: Clarendon Press.

Sklar, L. 1995. *Physics and Chance: Philosophical Issues in the Foundations of Statistical Mechanics*. London: Cambridge University Press.

Smith, A. 1976. *An Inquiry into the Nature and Causes of the Wealth of Nations*, vol. 2. *The Glasgow Edition of the Works and Correspondence of Adam Smith*. London: Oxford University Press.

———. 1976. *The Theory of Moral Sentiments*, ed. D. D. Raphael and A. L. Macfie. Oxford: Clarendon Press.

Smith, E., and D. K. Foley. 2008. "Classical Thermodynamics and Economic General Equilibrium Theory." *Journal of Economic Dynamics and Control* 32(1): 7–65.

Smith, J. M. 1972. "Game Theory and the Evolution of Fighting." In *On Evolution*, ed. J. M. Smith, 8–28. Edinburgh: Edinburgh University Press.

———. 1974. "The Theory of Games and the Evolution of Animal Conflicts." *Journal of Theoretical Biology* 47(1): 209–21.

———. 1982. *Evolution and the Theory of Games*. London: Cambridge University Press.

Sorensen, T. 2008. *Counselor: A Life at the Edge of History*. New York: HarperCollins.

Souma, W. 2001. "Universal Structure of the Personal Income Distribution." *Fractals* 9(4): 463–70.

Spatscheck, C. 2012. "Creating Capabilities: The Human Development Approach." *European Journal of Social Work* 15(3): 413–15.

Stanley, H. E., L. A. N. Amaral, D. Canning, P. Gopikrishnan, Y. Lee, and Y. Liu. 1999. "Econophysics: Can Physicists Contribute to the Science of Economics?" *Physica A: Statistical Mechanics and Its Applications* 269(1): 156–69.

Stiglitz, J. E. 2011. "Of the 1%, By the 1%, For the 1%." *Vanity Fair*, March 31. http://www.vanityfair.com/news/2011/05/top-one-percent-201105.

———. 2012. *The Price of Inequality: How Today's Divided Society Endangers Our Future*. New York: Norton.

———. 2015. *The Great Divide: Unequal Societies and What We Can Do About Them*. New York: Norton.

Stratton, L. S. 2001. "Why Does More Housework Lower Women's Wages? Testing Hypotheses Involving Job Effort and Hours Flexibility." *Social Science Quarterly* 82(1): 67–76.

Strom, S. 2016. "At Chobani, Now It's Not Just the Yogurt That's Rich." *New York Times*, April 26.

Theil, H. 1967. *Economics and Information Theory*. Amsterdam: North-Holland.

Thomasson, E. 2013. "Swiss Back Executive Pay Curbs in Referendum." *Reuters*, March 3. http://www.reuters.com/article/us-swiss-regulation-pay-idUSBRE92204N20130303.

Thomson, W. 2007. "Fair Allocation Rules." Working Paper. Rochester, NY: University of Rochester Center for Economic Research.

Tolman, R. C. 1938. *The Principles of Statistical Mechanics*. North Chelmsford, MA: Courier Corporation.

Tomasi, J. 2012. *Free Market Fairness*. Princeton, NJ: Princeton University Press.
Tribus, M., and E. C. McIrvine. 1971. "Energy and Information." *Scientific American* 225(3): 179–88.
Turnbull, H. W. 1959. *The Correspondence of Isaac Newton: 1661–1675*. London: Cambridge University Press.
van Eijck, J. and A. Visser. 2012. "Dynamic Semantics." *The Stanford Encyclopedia of Philosophy* (Winter 2012 Edition), ed. E. N. Zalta. https://plato.stanford.edu/archives/win2012/entries/dynamic-semantics/.
Venkatasubramanian, V. 2007. "A Theory of Design of Complex Teleological Systems: Unifying the Darwinian and Boltzmannian Perspectives." *Complexity* 12(3): 14–21.
—. 2009. "What Is Fair Pay for Executives? An Information Theoretic Analysis of Wage Distributions." *Entropy* 11(4): 766–81.
—. 2010. "Fairness Is an Emergent Self-Organized Property of the Free Market for Labor." *Entropy* 12(6): 1514–31.
—. 2011. "Systemic Failures: Challenges and Opportunities in Risk Management in Complex Systems." *AIChE Journal* 57(1): 2–9.
Venkatasubramanian, V., S. Katare, P. R. Patkar, and F.-P. Mu. 2004. "Spontaneous Emergence of Complex Optimal Networks through Evolutionary Adaptation." *Computers & Chemical Engineering* 28(9): 1789–98.
Venkatasubramanian, V., Y. Luo, and J. Sethuraman. 2015. "How Much Inequality in Income Is Fair? A Microeconomic Game Theoretic Perspective." *Physica A: Statistical Mechanics and Its Applications* 435: 120–38.
Venkatasubramanian, V., D. N. Politis, and P. R. Patkar. 2006. "Entropy Maximization as a Holistic Design Principle for Complex Optimal Networks." *AIChE Journal* 52(3): 1004–09.
Venkatasubramanian, V., and Z. Zhang. 2016. "Tecsmart: A Hierarchical Framework for Modeling and Analyzing Systemic Risk in Sociotechnical Systems." *AIChE Journal* 62(9): 3065–84.
von Helmholtz, H. 1883. *Wissenschaftliche Abhandlungen*, vol. 2. Leipzig: JA Barth.
Waldron, J. 2001. "What About Bert?" *London Review of Books* 23(15).
Walker, P. 2012. "A Chronology of Game Theory." http://www.econ.canterbury.ac.nz/personal_pages/paul_walker/gt/hist.htm.
Weissmann, J. 2014. "Americans Have No Idea How Bad Inequality Really Is." *Slate*, September 26. http://www.slate.com/articles/business/moneybox/2014/09/americans_have_no_idea_how_bad_inequality_is_new_harvard_business_school.html.
Wenar, L. 2013. "John Rawls." In *Stanford Encyclopedia of Philosophy*. https://plato.stanford.edu/entries/rawls/.

Willis, G. 2011. *Wealth, Income, Earnings and the Statistical Mechanics of Flow Systems*. Technical report. MPRA.

Willis, G. and J. Mimkes. 2004. "Evidence for the Independence of Waged and Unwaged Income, Evidence for Boltzmann Distributions in Waged Income, and the Outlines of a Coherent Theory of Income Distribution." Arxiv preprint, https://arxiv.org/abs/cond-mat/0406694.

Wolpert, D. H. 2006. "Information Theory—the Bridge Connecting Bounded Rational Game Theory and Statistical Physics." In *Complex Engineered Systems*, ed. D. Braha, A. Minai, and Y. Bar-Yam, 262–90. New York: Springer.

Wright, E. O. 2010. *Envisioning Real Utopias*, vol. 98. London: Verso.

Yakovenko, V. M. 2012. "Statistical Mechanics Approach to the Probability Distribution of Money." In *New Approaches to Monetary Theory: Interdisciplinary Perspectives,* ed. P. Bridel, 104–23. London: Routledge.

Yakovenko, V. M., and J. B. Rosser, Jr. 2009. "Colloquium: Statistical Mechanics of Money, Wealth, and Income." *Reviews of Modern Physics* 81(4): 1703.

Zabojnik, J. 1996. "Pay-Performance Sensitivity and Production Uncertainty." *Economics Letters* 53(3): 291–96.

Zillman, C., and S. Jones. 2015. "Seven Fortune 500 Companies with the Most Employees." *Fortune Magazine*.

# Index

Ackerman, F., 131–33, 135
aggregate income share data, 196
Ahituv, A., 51
AI. *See* artificial intelligence
Akerlof, G. A., 52
American capitalist society, economic ecosystem in, 185
Ananda Malai confections fairness example: fairest income distribution scenario in, 94–96; known aggregate demand scenario in, 92–94, 93; lock-and-key market-company relationship in, 96–97; overview, 89–90; unknown demand scenario in, 90–92
Apple, 178, 188
arbitrariness: avoidance principle of, 88; metric, 91; structural, 91
Ariely, Dan, 9–11
Arrow-Debreu model, 134–35
*Ars Conjectandi* (Bernoulli, J.), 89
artificial intelligence (AI), xviii
asymptotic stability, 121–23, 129, 139
Atkinson, A. B., 138, 235
atomic theory, 139
Australia: Gini coefficients for, *159–60*; income inequality in, *150*, 155; maximum income in, *146*; pay ratios in, 174
authoritarian society, 206
Avogadro number, 70
avoidance principle, of arbitrariness, 88

Bebchuk, Lucian, 6–7; on CEOs pay, 170, 174, 177; on rent-seeking behavior, 45
Bentham, Jeremy, 27; on goal of life, 36; insights of, 141; SO and, 118; on societal happiness, 196, 211
Bernoulli, Daniel, 227
Bernoulli, Jacob, 89
*BhuVai*, 24, 130; as envy-free society, 39; equal happiness in, 39, 43, 212; equilibrium income distribution in, 84–86; fair capitalist society and, 38–41; homogenous population in, 212; as hybrid utopia, 192–93, 195; income distribution in, 44, 144; income inequality in, 147; lognormal distribution of, 154; real-world societies and, 143,

*BhuVai* (*continued*)
195–96. *See also* distributive justice, in *BhuVai*
bias, 81; least biased distribution, 99, 101
Boltzmann, Ludwig, 62, 65, 66, 70; entropy and, 74, 104, 211, 224. *See also* Boltzmann-Gibbs distribution; Clausius-Boltzmann magnitudes
Boltzmann constant, 62, 70, 84
Boltzmann distribution, 69, 83; exponential distribution and, 85, 229
Boltzmann-Gibbs distribution, 114, 207, 228, 235
bottom-up perspective, 55
Britain, pay ratios in, 174
Buffett, Warren, 169, 178
Bureau of Labor Statistics, 184

Canada: Gini coefficients for, *159–60*; income inequality in, *152*, 155; maximum income in, *146*; pay ratios in, 174
canonical ensemble, 68
*Capital in the Twenty-First Century* (Piketty), 2
capitalism, 238; ersatz, 194, 237; extreme income inequality threatening, 15; income distribution and, 128; protecting, 129; Smith on, 116; unfairness and, 125. *See also* fair capitalist society
Cardano, Gerolamo, 88–89
career advancement, motivation for, 214–16
Carnot, Sadi, 65; as father of thermodynamics, 74
Cassidy, John, 2
CEOs: company success due to, 175, 176–77; compensation metrics relating to, 177; excessive compensations for, 45, 169, 181–82; as hired hand, 7; income percentage of, 177–81; measurable accomplishments of, 176; new generation of, 177; as sports/movie stars, 175–77; stakeholders relating to, 183; value of, 169–70, 177–81
CEOs, pay ratios of, 5–7, *6*, 24, 169, 178; actual, estimated, *8*; in Switzerland, 161
CEOs pay: Bebchuk on, 170, 174, 177; Fried on, 170, 174, 177; Reich on, 177; reining in of, 181–84; Stiglitz on, 177
Chebychev inequality, 100
chemical engineering, xvii–xviii
chemical potential, 85
Chesterton, G. K., 169
child labor, 125
Clausius, Rudolf, 65, 66; on entropy, 73–74, 211
Clausius-Boltzmann magnitudes, 22
closed thermodynamic system, 68
compensation: excessive, 45, 169, 181–82; overcompensation, 171. *See also* CEOs pay; executive pay; fair pay; pay packages
compensation metrics, 177
competitive interaction, 51
*Concerning Human Understanding* (Locke), 37
congestion games, 57–61
*Conscious Capitalism* (Mackey and Sisodia), 183, 189
Constitution, U.S., 32, 35, 129
control, extreme income inequality and, 13–14
cooperative games, 56
corporate governance, 236

cost-benefit trade-offs: home-buying example, 217–20; job-offers example, 221–23
Crystal, Graef "Bud," 7, 177

Darwin, Charles, 80, 207–8, 211
Darwinian evolution, statistical teleodynamics and, 207–8
Declaration of Independence, 32, 35
Denmark, 236; income inequality in, *149*, 154, 165; maximum income in, *146*; minimum wage for, 145
desert view of fairness, 98–102
design, 81, 202–4, 206, 207; and control principles, for fair capitalist society, 30; of hybrid societies, 192–98; perspective of, 102
design postulate, maximum fairness as, 204–5, 206, 207, 210
design principle, maximum fairness as, 192–98, 200, 246*n*1
difference principle, 28, 32, 192
distribution: maximum, 109–10, *110*; of wealth, *11*, *12*. *See also specific distribution*
distributive justice, xxi; addressing challenge of, 141; Dworkin on, 29, 193; happiness and, 41; Rawls on, 28, 193, 194
distributive justice, in *BhuVai*: equilibrium income distribution and, 61–62; game theory formulation and, 55–61; ideal *versus* real-life free market in, 44–48; income distribution and, 44; net utility's inverted u-profile in, 51–54, *53*; overview, 43; restless agents model and, 49–55; statistical thermodynamics connection, 62–63
Drucker, Peter, 7, 177

Dworkin, Ronald, 22, 24, 26, 27; on distributive justice, 29, 193; on envy-free society, 196; on equality of welfare, 39–40; framing fair society, 31; insights of, 141; mathematization of, xxi; political economic theory of, 192; resource egalitarianism of, 29; rigor of, 33; thought experiment of, 29–30
dynamics: evolutionary, 112; of free-market society, xix; replicator, 112–15. *See also* free-market dynamics
dynamic systems, 30, 65, 81

economic ecosystem, 189; in American capitalist society, 185; companies in, 185; customer with, 187; enabling elements of, 186; invaluable role of, 185; investor in, 187–88; market for, 187; nurtured and support of, 186; recognition and reward of, 187; in Silicon Valley, 186, 188; systemic failure of, 187; systemic success of, 186; technologies with, 187
economic growth: of India, 225–26; innovation relating to, 227; per capita GDP relating to, 225–26; of United States, 225–27
economic prescriptions, 235
"Economics Is Not a Morality Play" (Krugman), 115
economy, 234; entropy in, 21–22; inequality of, 8–12; laissez-faire, 132; in Scandinavia, 166; in Soviet Union, 166
econophysicists, 227–30
econophysics, 227–30; models of, 20–22

econophysics, economics and, 228; fairness relating to, 228; zero intelligence relating to, 228
education and skills assistance, 236
efficiency: robustness and, xviii, 36, 126–28; systems engineering on, 127–28
egalitarianism, 22; pragmatic, 33, 193; of Rawls, 29–30, 195; resource, 29
Einstein, Albert, 65, 191, 199
empirical studies, 233–34
employment, reasons for seeking, 49–50
entropy: Boltzmann and, 74, 104, 211, 224; Clausius-Boltzmann magnitudes of, 22; Clausius on, 73–74, 211; correcting misinterpretations, 102–12; discovery of, 66; in economics, 21–22; econophysics, economics and, 228; fairness and, xix, 89–105; in free-market dynamics, 103, 105–8; in free-market society, xxii; as gem, xxi; Gibbs on, 66, 104, 211; Helmholtz on, 104; introduction to, 24; invisible hand and, 108–11; law of, 70; lognormal distribution and, 162; maximizing, 62; maximum distributions, 109–10, *110*; principle of maximum, 101–2, 109, 111, 205, 224, 235; in qualitative theory of fairness, 23; rediscovery of, 104; robustness and, 105; in statistical thermodynamics, 69–77; as systems property, 75. *See also* maximum entropy design postulate
envy-free society, 39, 196

*equal a priori* probability postulate, 76, 194, 209
equal happiness, 39, 43, 212
equality: of access, 51; in hybrid societies, 193; in input, 200; in opportunity to compete, 200; in outcome, 200; principles of, 27–28, 87, 193, 195, 198, 199, 200; of resources, 29–30, 39
equality of welfare, 29, 39–40; at equilibrium, 117
equilibrium: equality of welfare at, 117; in game theory, 55–57; social optimality at, 118–20; statistical, 69; statistical mechanical, 62; strategic, 56; thermodynamic, 69, 71, 229; utility at, 234. *See also* Nash equilibrium; statistical equilibrium; Walrasian general equilibrium model
equilibrium density curve, 136–37
equilibrium distribution: energy, 84; fairness of, 86–102; income, 61–62, 84–86, 121–23; statistical thermodynamics and, 65–77
equilibrium energy distribution, 84
equilibrium income distribution, 61–62, 84–86, 121–23
equilibrium postulate, 209
equivalence value, 73
*Essai philosophique sur les probabilités* (Laplace), 89
Europe, income inequality in, *4*
evolutionary dynamics, 112
executive pay, 5–7
exercising, 53
exponential distribution, 85, 229
exponential stability, 129
extra consumption, 156
extreme income inequality, 155–57;

capitalism threatened by, 15; cause of, 178; consequences of, 14-15; control and, 13-14; moral issue of, 13, 16; stability and, 15; as systemic failure, 237-38; in United States, 238; as worrisome, 12-16
extrinsic properties, 68, 81-82

factory labor, 124-25
fair capitalist society: *BhuVai* and, 38-41; design and control principles for, 30; new theory of hybrid society, 31-34; philosophical perspectives, 25-30; purpose of society in, 34-38; theoretic perspective, 30-34
fair competitive environment, 51
fair distribution of income, 191
fair equality of opportunity, 28, 51, 197
fair equality of opportunity principle, 194-95
fairest income distribution, 217; scenario, 94-96
fairest inequality of income, 234
fair-market hypothesis, 108-11, 235
fairness: Ananda Malai confections example, 89-97; arbitrariness avoidance principle in, 88; desert view of, 98-102; econophysics, economics and, 228; entropy and, xix, 89-105; equality principle in, 87; of equilibrium distribution, 86-102; fair equality of opportunity, 28; in France, Germany, Switzerland, 147; of free-market society, 123-30; Gini coefficient and, 158; intercategory, 216; intracategory, 216; job categories and, 96-97; literature on, 25-26; lognormal distribution and, xix, 23-24, 143; maximizing, 127; moral principles of, 26; multiplicity and, 89-102; new mathematical theory of, 16-24; new measure for income inequality, 147; in pay ratios, 47-48; principle of maximum fairness, 33, 140, 200; principles of, 87-89; proportionality principle in, 87-88; quantitative theory of, 22-24; in United States, 145-47; utility function models of, 216-17. *See also* free-market fairness restoration; maximum fairness; theory, of fairness in economics and political philosophy; unfairness
fair opportunity, at better careers, 216
fair pay, for executives, 169-70; company success relating to, 175-89; theory's predictions, for large corporations, 171-74
fair shot, 49
Federal Reserve Board, 127
Feynman, Richard, 79, 140
"Financial Crisis, The: Why Have No High-Level Executives Been Prosecuted?" (Rakoff), 181
Financial Crisis Inquiry Commission, 181
Fisher, F. M., 131, 135
fitted power law, *165*
Fleischacker, Samuel, 116
*Fortune*, 7
four-class societies, 196
framing fair society, 31
France: fairness in, 147; Gini coefficients for, *159-60*; income inequality in, *151*, 155; maximum income in, *146*; pay ratios in, 174
Freeland, Chrystia, 11

free-market based hybrid societies, 34, 236
free-market capitalist societies, political philosophy for, 191
free-market complexity model, 217–25, 247n4; cost-benefit trade-offs, home-buying example, 217–20; cost-benefit trade-offs, job-offers example, 221–23; equations relating to, 217; principle of maximum entropy, 224; utility function features, 223–24
free-market dynamics, 98, 228; CEO value and, 169–70, 177–81; entropy in, 103, 105–8; formulation of, 55–61
*Free Market Fairness* (Tomasi), 30
free-market fairness restoration, 241
free-market principles, of Nozick, 32, 192, 194, 211, 235
free-market society, xxii, 123–30, 191–92, 234; characteristics of, 43; dynamics of, xix; economist's perspective, xx; entropy in, xxii; fairest distribution of income in, 217; fairness, stability, robustness of, 123–30; foundational elements of, 130–31, 239; Gedankenexperiment in ideal, 46–49; hybrid society as, 34, 236; ideal *versus* real-life, 44–48; justification for, 138; moral issue of, 17–18; Nozick on, 117; pay distribution in ideal, 46; Reich on, 45, 116, 170, 201, 241; stability of, 20, 184; Stiglitz on, 116; study of, xvii–xviii; in United States, 166–67, 237, 247n5. *See also* ideal free-market society
Freud, Sigmund, xxi–xxii
Fried, Jesse, 6–7; on CEOs pay, 170, 174, 177; on rent-seeking behavior, 45

Galileo, xvii
games. *See specific games*
game theoretic-statistical mechanics perspective, 208
game theory, xix, xxii, 24; formulation of free-market dynamics, 55–61; players, strategies, equilibrium, 55–57; population games and potential function, 57–61; potential function in, 108–9, 211, 234, 235; statistical mechanics and, 229
gauge symmetry, 199
Gedankenexperiment, 46–49
*General Theory, The* (Keynes), 106
George, Bill, 183–84, 189
Georgescu, Peter, 15
Germany: fairness in, 147; Gini coefficients for, *159–60*; income inequality in, *151*, 155; maximum income in, *146*; pay ratios in, 174
Gibbs, Josiah Willard, 65, 68, 227; on entropy, 66, 104, 211. *See also* Boltzmann-Gibbs distribution
Gini coefficients, 24; early to mid-2000s, *159–60*; fairness and, 158; income equality global trends, *159–60*; lognormal distribution compared to, 158–62
Global Financial Crisis of 2007-2009, 134, 187
goal-driven systems, 31
Golden Rule, 140
grand canonical ensemble, 68

Hamilton, C. V., 37
happiness: distributive justice and, 41; equal, 39, 43, 212; measuring,

40; pursuit, 37–38, 211; societal, 21, 196, 211; *World Happiness Report*, 40. *See also* utility
hedge funds, 53
Helmholtz, Hermann, 65; on entropy, 104
Helmholtz free energy, 85, 114
Hilbert, David, xx
Hill, Christopher T., 199
*Homo Economicus* model, 209
homogenous population, in *BhuVai*, 212
"How Should We Use Entropy in Economics?" (Jaynes), 105
hybrid avatars, 202
hybrid societies, 246n2; design of, 192–98; as free-market based, 34, 236; independence in, 34; liberty and equality in, 193; new theory of, 31–34; systems engineering perspective of, 34–38; utopian, 33, 38–41. *See also BhuVai*

ideal free-market society, 44–48, 191; equality of welfare at equilibrium in, 117; moral justification for, 115–21; NE and SO in, 120–21; pay distribution in, *46*; Smith on, 119, 194, 211, 236, 237, 241; SO at equilibrium in, 118–20
ideal income distribution, 138
ideal lognormal distribution, 145–47, 170
ideal pay ratios, for CEOs, *8*
inalienable properties, 68
income distribution, 1; in *BhuVai*, 44, 144; capitalism and, 128; correcting misinterpretations of entropy, 102–12; fairness, stability, robustness of free-market society and, 123–30; fairness of equilibrium distribution, 86–102; ideal, 138; intrinsic and extrinsic properties of statistical teleodynamics, 81–82; lognormal distribution for, 138, 193, 236; model predictions, *146*; moral issue of, 17–18; moral justification of ideal free-market society and, 115–21; Nozick on, 44; phase space in teleodynamic systems and, 82–84; power and, 170; principles of fairness and, 87–89; questions for, 17–20; replicator dynamics and, 112–15; stability of, 129; stability of equilibrium income distribution in, 121–23; teleogical systems and, 79–81; Walrasian general equilibrium model and, 130–35, 139. *See also* equilibrium income distribution; fairest income distribution
income equality, Nozick on, 192, 193
income inequality, xvii, 201; American and Swedish preference for, 9, *10*; Anglo-Saxon, 1910–2010, *4*; in Australia, *150*, 155; in *BhuVai*, 147; in Canada, *152*, 155; comparing theory with reality, 144, *145*; in Denmark, *149*, 154, 165; economic inequality and, 8–12; Europe, 1910–2010, *4*; in executive pay and CEO pay ratios, 5–7, *6*, 24; in France, *151*, 155; in Germany, *151*, 155; Gini coefficient *versus* lognormal distribution in, 158–62; identifying, 2; in Japan, *152*, 155; loss of growth due to extreme inequality, 155–57; in Netherlands, *150*, 155; new mathematical theory of fairness,

income inequality (*continued*)
  16–24; new measure of fairness
  in, 147; in Norway, *148*, 154,
  165; in one-class and two-class
  societies, 162–65, *163*, 196,
  212–13; overview, 143–44;
  qualities of rich and, 157–58; rise
  of top 1% and fall of bottom 90%,
  1–5; in Scandinavia, 157–58;
  summary, 165–67; in Sweden,
  *148*, 154, 165; in Switzerland,
  *149*, 154, 165; unfairness in, 158,
  165; in United Kingdom, *153*,
  155; when needed, 12. *See also*
  extreme income inequality
income inequality, in United States:
  1910–2010, *2*; 1930s to 2010s,
  *153*, 155–57; 1945–1975,
  155–57; top 1%, 1910–2010, *3*
income percentage, of CEOs,
  177–81
income shares, 236
India, economic growth of, 225–26
indifference, principles of, 89, 210
inequality: of economy, 8–12;
  Jensen's, 100; wealth of, *5*. *See
  also* extreme income inequality;
  fairest inequality of income;
  income inequality; income
  inequality, in United States
inheritance tax, 236
*Inquiry into the Causes of the Wealth
  of Nations, An* (Smith), 18
*In Search of Excess* (Crystal), 7
insights, 141; conceptual, 140; for
  cost-benegit trade-offs, 222–24
intercategory fairness, 216
intracategory fairness, 216
intrinsic properties, 81–82
inverted u-profile, 51–54, *53*, 223
investor, in economic ecosystem,
  187–88

invisible hand: entropy and, 108–11;
  mathematical basis for, 132;
  Nozick and, 87; of Smith, xix,
  18–20, 48, 234–35; in *The Theory
  of Moral Sentiments*, 18–19; in
  *Wealth of Nations*, 19
isolated thermodynamic system, 68

Japan: Gini coefficients for, *159–60*;
  income inequality in, *152*, 155;
  maximum income in, *146*; pay
  ratios in, 174
Jaynes, E. T., xxi–xxii, 66, 70,
  88–89, 105, 208, 211
Jensen's inequality, 100
jobs, 39, 54, 96–97
Jobs, Steve, 99, 178, 189
Johnson, Samuel, 36, 211
Joule, James Prescott, 65
*Justice* (Sandel), 115
*Justice as Fairness* (Rawls), 27
Justice as Fairness theory, 87

Kapur, J. N., 110
Kelvin, Lord, 65
Kenworthy, Lane, 16
Keynes, John Maynard, xvii, 89, 106,
  128
Kiatpongsan, Sorapop, 8
Kirman, Alan, 134–35, 227
Kleinberg, J., 61
known aggregate demand scenario,
  92–94, *93*
Krugman, Paul, 115

labor, dignity of, 212
Lagrangian, 60, 85–86, 101
laissez-faire economy, 132
Langone, Ken, 15
Laplace, Pierre Simon, 89
law of conservation of energy,
  66, 70

least biased distribution, 99, 101
Lederman, Leon M., 199
Lewis, C. S., 37
lexical ordering, 28
*Liber de Ludo Aleae* (Cardano), 88–89
libertarianism, 22, 30, 33, 193, 195, 196
liberty principle, 27–28, 193, 195, 211
Lincoln, Abraham, 1
LIS. *See* Luxembourg Income Studies Database
local competition, 50
lock-and-key market-company relationship, 96–97
Locke, John: insights of, 141; mathematization of, xxi; on pursuit of happiness, 37–38, 211
lognormal distribution, 133, 212, 230; of *BhuVai*, 154; entropy and, 162; fairness and, xix, 23–24, 143; Gini coefficient compared to, 158–62; as gold standard, 141; ideal, 145–47, 170; for income distribution, 138, 193, 236; misidentifying, 163–66; misinterpreting, 22; overlapping, 196; pay and salaries based on, 170–71; proving, 85–87, 101–2, 117, 137; theory's predictions, for large corporations with, 171
lottery effect, 11
Lucas, Robert E., Jr., 26, 225
Luxembourg Income Studies Database (LIS), 159, *159*, *160*
Lyapunov, Aleksandr Mikhailovich, 122–23, 212, 217
Lyapunov function, 122–23, 235

Mackey, John, 183
macrostates: extremely likely and unlikely, *74*, 75; in statistical thermodynamics, 67–69
macro survival postulate, 201–2, 206, 207
mathematical basis, for invisible hand, 132
mathematical foundations, of a utopian capitalist society, 192–201
mathematical framework, for theory examination, 182
mathematical theory, of fairness, 16–24
mathematical tools, 135, 140
mathematization, xxi
maximum distribution, entropy, 109–10, *110*
maximum entropy design postulate, 204–5, 206, 207, 210
maximum entropy principles, 101–2, 109, 111, 205, 224, 235
maximum fairness, 33, 140; in design postulate, 204–5, 206, 207, 210; as design principle, 192–98, 200, 246n1; maximum entropy principle and, 235; as symmetry principle, 198–201, 247n3; as universal principle, 200
maximum income, *146*
maximum uncertainty, 204, 207
Maxwell, James Clerk, 65
measurable accomplishments, of CEOs, 176
measure of uncertainty, 70, 210
metric arbitrariness, 91
microcanonical ensemble, 68
microstates: making additional, 75; in statistical thermodynamics, 67–69
micro survival postulate, 208–9
Mill, John Stuart, 24, 211; framing fair society, 31; on goal of life, 36;

Mill (*continued*)
  insights of, 141; political economic theory of, 192; SO and, 118; on societal happiness, 21, 196; utilitarianism of, 27, 34
minimal state argument, 28–29
minimum wage, 34, 44, 145, 236
Minsky, H. P., 135
Minsky moment, 134
Mirowski, Philip, 106, 134–35, 227
molecular society, 67–69
molecules, 71, *72, 73, 74*
Monderer, D., 57–58, 108
moral issues, 13, 16, 17–18, 26; with ideal free-market society justification, 115–21
Morgan, J. P., 7, 170, 178
Morgenstern, Oskar, 55–56
multiplicity: fairness and, 89–102; increasing, 74; in statistical thermodynamics, 67–69
Mylan NV, 179, 180

*Narasimha Avatar*, 192, 195
Nash, John, 56–57
Nash equilibrium (NE), 56–57, 59, 62, 84, 118–19, 217; arriving at, 139, 210; in ideal free-market society, 120–21; indifference with, 210; mixed strategies with, 209–10; statistical equilibrium and, 209–10, 211, 229
National Bureau of Economic Research, 184
naturally arising distribution, Nozick on, 110–11
Navier-Stokes equations, 139
NE. *See* Nash equilibrium
negative definite, 123
negative semidefinite, 123
Netherlands: average salaries in, 145; Gini coefficients for, *159–60*; income inequality in, *150*, 155; maximum income in, *146*
net utility's inverted u-profile, 51–54, *53*
Newton, Isaac, 191
Newton's laws of motion, 68, 70–73; for molecular collisions, 84; prisoners of, 79–80
*New York Times*, 15
Noether, Emily, 199
Noether's theorem, 199
noncooperative games, 56
nonidealities, 192
normative analytical theory, 191, 234
Norton, Michael, 8–11
Norway: Gini coefficients for, *159–60*; income inequality in, 148, 154, 165; maximum income in, *146*; minimum wage for, 145
Nozick, Robert, 22, 24, 26, 27; framing fair society, 31; free-market principles of, 32, 192, 194, 211, 235; on free-market society, 117; on income distribution, 44; on income equality, 192, 193; independence and, 34; insights of, 141; invisible hand and, 87; libertarianism of, 30, 195, 196; mathematization of, xxi; minimal state argument of, 28–29; on naturally arising distribution, 110–11; purpose of society and, 36–37; rigor of, 33
Nussbaum, M. C., 22, 33, 34

OECD. *See* Organisation for Economic Co-operation and Development
one-class societies, 162–65, *163*, 196, 212–13
open thermodynamic system, 68

operational perspective, 102
optimally design, 81
optimally robust design, 81; postulate, 202–4, 206, 207
Organisation for Economic Co-operation and Development (OECD), 159
Ormerod, P., 22, 230
overcompensation, 171

Pareto distribution: misidentifying, 133; as power law, 162–63, 171
Pareto efficiency, 109, 204, 205
Pareto-optimal outcome, 119, 131, 202
pay and salaries, based on lognormal distribution, 170–71
pay compression policy, 182–83
pay distribution, in ideal free-market society, 46
payoff, 55, 57, 111; personal attributes and preference of, 213–14
pay packages, 172–74, *173*, 246*n*1
pay ratios, 172–73, *173*; actual, estimated, ideal, *8*; in Australia, 174; in Britain, 174; in Canada, 174; for CEOs, 5–7, *6*, *8*, 24, 161, 169, 178; in executive pay, 5–7; fairness in, 47–48; in France, 174; in Germany, 174; in Japan, 174; for Switzerland CEOs, 161; underestimation of, 9, 10; in United States, 174
*Pay without Performance: The Unfulfilled Promise of Executive Compensation* (Bebchuk and Fried), 6–7
per capita GDP, 225–26
performance: levels of, 80; peak, 215
performance target, 171–72; postulate, 202, 206, 207

Perline, R., 22, 230
phase space: defined, 69; evolving systems and, 70; in statistical thermodynamics, 69–77; in teleodynamic systems, 82–84
Piketty, Thomas, 2, 45, 138, 144
Planck, Max, 65
Poincaré, Henri, 224
political economic theory: of Dworkin, 192; of Mill, 192
political philosophy, 191, 234
population games, 57–61
potential function, 57–62, 108–9
potential game theory, 57–61
power, income and wealth with, 170
pragmatic egalitarianism, 33, 193
pragmatic libertarianism, 33, 193
predistribution, 156
preference parameters, 223
probability theory, 55
progressive taxation, 236
proportionality principle, 87–88, 98
punishment, 88
pure model societies, 192
purpose: of corporation, 240; of life, 240; of society, 34–38, 239–40
pursuit of happiness, 37–38, 211

qualitative insights, 140
qualitative theory of fairness: entropy in, 23; outline of, 22–24; predictions of, 23
quantitative theory, 22–24, 192, 196

Rakoff, J. S., 181
randomness, 104–5
Rankine, William, 65
Rawls, John, 22, 24, 26, 235; difference principle of, 32, 192; on distributive justice, 28, 193, 194; egalitarianism of, 29–30, 195; fair equality of opportunity

Rawls (*continued*)
  principle of, 194–95; framing fair society, 31; on income, 48; insights of, 141; Justice as Fairness theory of, 87; mathematization of, xxi; minimum wage and, 34; principles of liberty and equality, 27–28, 193, 195, 211; purpose of society and, 36–37; rigor of, 33; strong state argument of, 28; veil of ignorance and, 9, 27
real-life nonideal conditions, 158, 192
real-world societies, 143, 195–96, 236
Reich, Robert, 5–6, 14–15, 107, 115; arguments of, 121, 205; CEO list by, 179; on CEOs pay,
Reich (*continued*)
  177; on free-market society, 45, 116, 170, 201, 241; on predistribution, 156
Reif, F., 69
rent-seeking behavior, 45
replicator dynamics, 112–15
resource egalitarianism, 29
restless agents model: distributive justice in *BhuVai* and, 49–55; utility of fair opportunity for better future, 50–51
Reuters, Thomson, 11
robustness: efficiency and, xviii, 36, 126–28; entropy and, 105; of free-market society, 123–30; maximizing of, 92–93, 97; optimally robust design, 81; stability and, 125–26; systems engineering on, 127–28
Roosevelt, Theodore, 26, 143, 188
Rosenthal, R. W., 57–58
rotational symmetry, of circle, 198–99

Saez, E., 138
salary scenarios, 172
Samuelson, Paul, 21–23, 108
Sandel, Michael, 115
Sandholm, W. H., 57, 61
*Saving Capitalism* (Reich), 5–6
say-on-pay policy, 181, 182
Scandinavia, 236–37; economy in, 166, 192; income inequality in, 157–58
Scanlon, Tim, 13–14, 115–16; arguments of, 121, 205
Schmidtz, D., 33
Schoenberg, E., 117
"Science and Method" (Poincaré), 224
Securities and Exchange Commission, 127
self-organized complex adaptive systems: design of, xviii; understanding, xvii
semiquantitative impressions, of market, 220
Sen, Amartya, 21, 26, 105; rigor of, 33; special treatment addressed by, 34
Sethuraman, Jay, xix
Shannon, Claude, 66, 70, 104, 208, 211
Shapley, L. S., 57–58, 108, 211
Silicon Valley, 186, 188
Simon, Herb, 170, 188
Sisodia, Raj, 183
slavery, 124
SMD theorem, 131
Smith, Adam, 1, 108, 143, 184, 189, 217; on capitalism, 116; on ideal free-market society, 119, 194, 211, 236, 237, 241; insights of, 141; invisible hand of, xix, 18–20, 48, 234–35; mathematization of, xxi; on pursuing interests, 39

socially optimal (SO), 118–21
societal happiness, 21, 196, 211
societies. *See specific societies*
society, purpose of, 34–38, 239–40
Soviet Union, economy in, 166
stability: asymptotic, 121–23, 129, 139; of equilibrium income distribution in, 121–23; exponential, 129; extreme income inequality and, 15; of free-market society, 20, 123–30, 184; of income distribution, 129; robustness and, 125–26
stakeholders, CEOs and, 183
star network, 203
statistical equilibrium: of gas molecules, 72, 73, 74; NE and, 209–10, 211, 229; reaching, 69; in statistical thermodynamics, 69–77; uniform distribution of molecules with, 71
statistical mechanics, xviii, xix, 66, 208, 229, 234; central concepts of, xxii; econophysics and, 20; introduction to, 24; statistical mechanical equilibrium, 62
statistical teleodynamics, xix, 24, 31, 229, 234; Darwinian evolution and, 207–8; developing, 79; intrinsic and extrinsic properties, 81–82; key features of, 197–98; transdisciplinary synthesis, 211–12
statistical teleodynamics, macroview, 206–8; perspectives of, 201, 235; postulates of, 201–5
statistical teleodynamics, microview, 210–12, 235; growth relating to, 208; postulates of, 208–9; survival relating to, 208
statistical thermodynamics, 229, 234; distributive justice in *BhuVai* connection, 62–63; equilibrium distribution and, 65–77; history of, 65–67; microstates, macrostates, multiplicity in, 67–69; phase space, entropy, statistical equilibrium in, 69–77; predicting system-level properties and dynamic behavior, 65. *See also* statistical mechanics

Stiglitz, Joseph, 13, 15, 115, 194, 237; arguments of, 121; on CEOs pay, 177; on free-market society, 116; on rent-seeking behavior, 45
Stirling's approximation, 61, 94
strategic equilibrium, 56
strategic games, 55
strong state argument, 28
structural arbitrariness, 91
survival of fittest, 32
Sweden, 236; Gini coefficients for, 159–60; income inequality in, 148, 154, 165; maximum income in, 146; minimum wage for, 145
Switzerland: CEO pay ratios in, 161; fairness in, 147; Gini coefficients for, 159–60; income inequality in, 149, 154, 165; maximum income in, 146; minimum wage for, 145; referendum on CEO pay in, 182
symmetry, 191, 198–201, 247n3
*Symmetry and the Beautiful Universe* (Lederman and Hill), 199
symmetry principle, 198–201, 247n3
system-level properties, 65
systems engineering: on efficiency and robustness, 127–28; perspective of hybrid society, 34–38; questions posed in, 35–36, 38
systems property, entropy as, 75
system-theoretic rationale, 193

tatonnement process, 133
taxes, 45, 236; United States and, 161
technologies, with economic ecosystem, 187
teleodynamic game, 85
teleodynamics. *See* statistical teleodynamics
teleodynamic systems, phase space in, 82–84
teleogical systems: human-engineered, 80; income distribution and, 79–81; levels of performance in, 80
teleological agents, 229
Theil index, 21
theoretical and agent-based simulation studies, 230–33
theory, of fairness in economics and political philosophy, 191; fair distribution of income, 191; with ideal free-market society, 191; normative analytical theory, 191, 234
*Theory of Games and Economic Behavior, The* (von Neumann and Morgenstern), 55–56
*Theory of Moral Sentiments, The* (Smith), 18–19
theory's predictions, for large corporations: data-gathering
theory's predictions (*continued*) about, 184; general principles relating to, 184; lack of accountability with, 178–79; lognormal distribution with, 171; mathematical framework for examination of, 182; overcompensation with, 171; pay packages relating to, 172–74, *173*, 246n1; pay ratios relating to, 172–74, *173*; performance targets with, 171–72; salary scenarios for, 172
thermodynamic game, 62, 114; equilibrium energy distribution for, 84
thermodynamics: Carnot as father of, 74; equilibrium, 69, 71, 229; laws of, 70, 208; potential of, 85; systems, 68. *See also* statistical thermodynamics
three-class societies, 196, 213
Tomasi, John, 30, 33
top-down perspective, 55
transdisciplinary synthesis, of statistical teleodynamics, 211–12
translational symmetry, 199
Turing Pharmaceuticals, 179–80
two-class societies, 162–65, *163*, 196, 212–13
*Two Treatises on Government* (Locke), 37

unfairness, 123; capitalism and, 125; in income inequality, 158, 165; in United Kingdom, 165; in United States, 165
United Kingdom: average salaries in, 145; Gini coefficients for, *159–60*; income inequality in, *153*, 155; maximum income in, *146*; unfairness in, 165
United States: actual wealth distribution in, *11*, *12*; economic growth of, 225–27; fairness in, 145–47; free-market society in, 166, 237, 247n5; Gini coefficients for, *159–60*; maximum income, *146*; pay ratios in, 174; tax rates and, 161; unfairness in, 165
United States, income inequality in: extreme, 238; 1910–2010, *2*;

1930s to 2010s, *153*, 155–57; 1945–1975, 155–57; top 1%, 1910–2010, *3*
universal health care, 236
universal principle, of maximum fairness, 200
unknown demand scenario, 90–92
upward mobility, 49, 51
utilitarianism, 27, 34
utility, 48, 193; at equilibrium, 234; of fair opportunity for better future, 50–51; inverted u-profile, 51–54, *53*; from job, 39
utility function: features of, 223–24; parameters of, 212–17
utility function models, of fairness, 216–17
utopia, 192–201, 236. *See also* hybrid societies
utopian capitalist society, mathematical foundations of, 192–201

valuation bubble, 174
*Vanity Fair*, 13

veil: of ignorance, 9, 27; of randomness, 105
vile maxim remark, 237, 239
von Mayer, Julius Robert, 65
von Newmann, John, 55–56, 104

Walrasian general equilibrium model, 130–35, 139
Watt, James, 65
wealth: distribution of, *11*, *12*; inequality of, *5*; power and, 170
*Wealth of Nations* (Smith), 19, 116, 217
Weyl, Hermann, xx
winner-take-all culture, 188, 189
women's suffrage, 124
World Economic Forum of 2015, 2
*World Happiness Report*, 40
World Top Incomes (WTI), 144, 147
Wozniak, Steve, 188–89
WTI. *See* World Top Incomes

Yellen, J. L., 52
Yu Luo, xix

zero intelligence, 228